Praise for *Climate Justice, Climate Hope*

If you feel overwhelmed about the climate crisis and think there is little you can do, this book is for you! Malcom and mohaupt's dialogical (ad)venture is an empowering call to action designed to raise not only our consciousness but also our voices, hands, and feet, all in order to "grow the bigger we." The result is a story-based, biblically rooted, hope-filled, eye-opening, deeply personal, outwardly practical call for the healing of the world that God so loves. It is a book of blessing.

—**William P. Brown**, William Marcellus McPheeters Professor of Old Testament, Columbia Theological Seminary

Michael Malcom and abby mohaupt have been in the trenches of the climate struggle for a generation, and their blend of deep faith and impassioned commitment is as fresh and important now as ever. Faith communities can and must rise—and Michael and abby are helping light the way.

—**Fletcher Harper**, executive director, GreenFaith

Climate Justice, Climate Hope is a powerful call to honor our sacred connection to all life. This book weaves together compelling stories and moral arguments, urging us to dismantle systems that harm both humanity and the earth. It inspires a vision of an inclusive and caring world where we recognize our responsibility to heal our planet and each other.

—**Ife Kilimanjaro**, executive director, US Climate Action Network, and author of *Re-Membering Purpose*

This is no ordinary conversation. It is sacred witness. Two wise climate activists speak with the urgency of prophets and the tenderness of pastors. Their stories awaken the heart and equip us to act boldly, faithfully, and imaginatively in the face of our shared crisis.

—**Brian Sauder**, president and CEO, Faith in Place

In an astounding feat, *Climate Justice, Climate Hope* enables readers to grapple with the role of racism and capitalism in the devastating realities of climate chansge, but without overwhelming us with despair. Malcom and mohaupt craft an accessible, fresh vision of environmentally just possibilities. They highlight stories of everyday business initiatives, neighborhood activism, and Christian faith resources that undermine planet-destroying, white-supremacist economic relationships and instead create bold, spiritually nurturing practices of equitable societal transformation.

—**Traci C. West**, James W. Pearsall Professor of Christian Ethics and African American Studies, Drew University Theological School; author of *Solidarity and Defiant Spirituality*

CLIMATE JUSTICE, CLIMATE HOPE

Foreword by Bill McKibben

CLIMATE JUSTICE, CLIMATE HOPE

BUILDING A MORAL ECONOMY

Michael Malcom and abby mohaupt

FORTRESS PRESS
Minneapolis

CLIMATE JUSTICE, CLIMATE HOPE

30 29 28 27 26 25 1 2 3 4 5 6 7 8 9

Library of Congress Control Number: 2025938610 (print)

Cover image: Abstract Impressionistic background by Oleg Agafonov via Getty Images
Cover design: Josh Eller

Print ISBN: 979-8-8898-3283-6
eBook ISBN: 979-8-8898-3284-3

CONTENTS

FOREWORD TO THE SERIES

Life's beauty, including simple human goodness, astounds. It feeds our souls. Yet, our economic lives—systems for producing, consuming, investing, fueling, and more—wage war with life on earth. The climate, water, soil, and air that we depend upon for life itself can no longer bear our extractive economies. Our economy may appear normal, but it is sick.

We can change course! Like a flock of birds swooping around to fly a different direction, we can redirect—from economies of exploitation, extraction, and fossil fuels to economies of well-being for all. In countless ways, ordinary people of all colors, creeds, and continents are living this vision into reality. Creative and courageous, they are students, business leaders, jobless people, scientists, lawyers, farmers, nonprofit workers, artists, religious leaders, factory workers, healthcare professionals, retirees, homemakers, educators, and more. Indeed, a splendid worldwide movement to heal economic life is unfolding.

These singers of hope say "no more" to economic practices that threaten earth's climate. "No more" to economic policies that drive some into anguishing poverty and plundering the earth, so that a few can amass great wealth and some can consume with abandon. In this book and the others in the series, you will meet these singers of hope and explore the spiritual roots of their courage, creativity, and commitment.

Will we step up to the plate? Take on this great journey of our time—the journey to forge ways of sharing life together that enable earth's life systems to thrive, and all people to have the necessities for life with justice and joy? We (the authors in this book series) invite you into—or more deeply into—this sacred journey of healing.

An amazing team of highly gifted authors has coalesced to create five small-books that, together with an introductory volume, provide invaluable guidance for the journey toward economies for life. The book in your hands is the first of those five. This powerful little volume highlights and builds on four frameworks that run throughout the series. The first is "three terrains of change"—behaviors, social structures, and mindsets—that interact and support each other. The second is "ten fingers on the hands of healing change," referring to ten different forms of action that work together, each opening doors to the others. The third is the idea that moving from exploitative fossil-fueled economies to life-giving economies is a sacred and healing journey. The final framework is the re-understanding of economies as webs of relationship. These frameworks are developed in the introductory volume of the series, *Building a Moral Economy: Pathways for People of Courage.*

All five small-books build on these frameworks. All highlight the role of religion and spiritualty in enabling the shift toward more equitable, ecological, and democratic forms of economic life. Yet each book has a unique voice and approach. *Climate Justice, Climate Hope* focuses more on the contributions of Christian traditions than do some of the subsequent books that look at multiple faith traditions. And this book is written by a team of two. They model the power and challenges of deep trusting collaboration between a Black person and a white person who share a profound commitment to facing the realities of climate change and its links to racism and exploitative economies with both searing honesty and inspiring hope and humor. In their hands, you will journey well!

As you read on, be sure to savor the wealth of resources on the website accompanying this book series. There you will meet a global community of companions on the journey. You will find also reading materials, films, websites, and other resources including liturgies, other rituals, art, worship resources, and more: https://buildingamoraleconomy.org/. Under the tab "Books" and then "Book 1: Climate," you will find resources particular to this book in the series.

Earth's song of hope resounds around the globe in movements to build economies—local, national, and global—that support the flourishing of all. If you are part of this visionary yet practical flock, may this book strengthen your engagement. If you are not, then welcome! May your trip through these pages bring you joyfully into this community of hope.

What a strange wonder to be alive at this turning point in history. Humans, by our actions, will shape the fate of life on earth. None before us have borne this moral weight. What we do matters! *It is, therefore, a good time to be alive.*

Cynthia Moe-Lobeda, series editor

FOREWORD

I got to read this powerful book for the first time in early February 2025, just as the full force of Trumpism was breaking across my country. The daily—hourly—headlines were all about domination—"energy dominance," taking over Greenland or Canada. The president circulated images of himself as a king wearing a crown. It was the logical extension, and also the reductio ad absurdum, of the forces in our society that the authors name so well: patriarchy, white supremacy, the baptizing of capitalism, and accumulation.

And at the same moment the climate was spinning ever more dangerously out of control: The great Los Angeles wildfires came in January 2025, the hottest January ever recorded, on the heels of the hottest year ever recorded. The only thing that separated those blazes from most of the climate disasters of the last decades was that it hit rich people as well as poor, wiping out whole neighborhoods of multi-million dollar homes.

It was, in other words, a dark time in our national and our planetary history, which made me all the more grateful for the deep argument contained in these pages. That argument, with which I agree, is that we still have some ability to shape the outcome of our perilous moment. But to do so we have to recognize our interconnectedness and revel in our solidarity, not our hyper-individual independence. I think there are many ways to make this real: We can, for instance, replace the centralized power stations with solar farms and wind turbines—because the energy of the sun and wind can't be hoarded or held in reserves, but instead is offered freely to us all.

Those technical changes, though, need to go hand in hand with a new attitude, one that rejects all that domination and cruelty, exchanging them for the pleasures of a life lived in connection with and in service to others. A life lived, in this case, in response to the Gospel command to love our neighbors. This is not an easy book—it speaks, and again I agree, in the language of sin. But it also shows the path toward some kind of redemption, and indeed toward a planet and a society that works for us all. I do not know if we will realize that vision—time is short, and the odds are steep. But I do know that trying to get there is the most important and most beautiful way to spend one's life. Consider this volume a guidebook.

Bill McKibben, Middlebury, VT, February 2025

INTRODUCTION

Letters to Our Children and to the Reader

To Malachi, Micah, Mareshah

IT FEELS WEIRD writing to you all knowing my words will live in this book. Writing makes my words more permanent than if you were hearing them from me. In addition, these words are laid bare for the entire world to gaze upon. Therefore, I must be responsible with the words that I impart to my future.

You all are my most prized possession. My greatest accomplishment is knowing that I have produced and raised the best parts of me to adulthood. Before your birth, I saw you. God showed me the benefit that you would bestow upon your mother and me. Your mother and I were young and trying to figure out life; however, we knew that we had to provide for you all. We put our heads down and we started our grind. We became beacons of capitalism—slaves to the grind.

I want to apologize to you all. Your mother and I were not great models of what God wants for our lives and society. I have learned that the constant grind that capitalism calls for has led to isolation and destruction. I failed to show you through our relationship, the importance of being inclusive. Not only is being inclusive important, it is beneficial.

My beloved offspring, being inclusive leads to growth and abundance. What we should have been showing you is that when we are inclusive we welcome other thoughts, customs, and cultures that enhance and grow us. As young Black adults, you know what it is like to feel oppressed and excluded. I remember the conversations that we experienced together during the 2020 uprisings. We spoke of what it felt like to witness George Floyd being lynched while crying out that he could not breathe. You have gone with me to speak with communities experiencing deep poverty and environmental racism.

The conditions that we have seen some of our fellow humans living in are due to an absence of community and care and to the soulless practices of a predatory economy. The words that I write in this book are a moral argument to evoke a moral response. This book is a call to be inclusive. It is a call to care. It is a call to help the people so that we can heal the planet. I am eternally grateful to rev. abby, PhD for providing me this platform and opportunity.

Love, peace, and light,
Daddy

Dear Juniper

When Rev. Michael and I were working on this book, you were too little to really understand the realities of climate change as a whole and how it connects to what we do with money.

But your life already has been affected by climate change. On a larger scale, we moved from Texas to Chicago because it was too hot in Texas in the summer for you to play outside because the temperatures are rising. We became sort of like climate migrants, though not at the level of risk that people have had to take in other parts of the world (such as Northern Africa and across Asia). On a much smaller scale, our dear Ed and Clare got you a sled for your third Christmas so you could play in the Chicago winter snow. You haven't been able to use it because there's less snow in winter than ever before. You're missing out on this simple joy. And yet—there is much more at stake in a world where climate injustice means that so many people go hungry, must leave their homes, face terrible destruction. As I write this, an inferno blazes through Los Angeles, affecting people we know and people we do not. Who knows what your generation will face?

When you were born, I was in the midst of finishing a dissertation on divestment from fossil fuel, and I was working with GreenFaith on climate finance stuff. That's how you first met Rev. Michael, outside of the White House when you were three months old and we called on then-President Biden to cease funding fossil fuels, following the call of Indigenous leaders. We must keep listening to their voices and teachings. We must keep showing up with courageous and faithful people on the front lines. I'm so grateful for this shared labor with Michael, that protest (and the ones before and after), and this book.

I don't know if what we (Rev. Michael and I) have done, in our lives and in this book, will be enough to change our economies to align with a more just world, a more climate-friendly world. Our family economy (yours and mine, Juniper) has been informed by our faith in Jesus, the one who toppled tables to demand a better economic system for all. We haven't always made all the right decisions. What I know is that we (you and I who are white and have so much privilege in this world) will keep showing up for the work, with God's help.

I love you, Juniper. It is for you that I do the work God calls me to.

With love,
your mama

Dear Reader

Early in our conversations about writing this book, we reminded each other that a moral economy that responds to climate change must dismantle white supremacy and capitalism as we know it. We're both pastors, rooted in activist communities. And so we looked at both the Bible and grassroots communities to guide us.

Two stories from the Bible grounded us as we sought to tell stories of an economy that is moral and hope-filled in a time of climate change. First, the story of Ruth and Boaz, rooted in gleaning practices and expectations to care for kin, has taught us about our responsibility (in God's economy) to care for people on the margins.[1] In a climate-changed world, this includes the whole ecology, but especially colonized and marginalized communities. Second, in the book of Acts, we see the abundance of God's economy.[2] The church began with people laying everything at the foot of the apostles so that none would be without. It was a profound way to care for one another, and it is a reminder that there is enough. Principles of capitalism have insisted that there is not enough and that everyone needs to compete for what there is. We (Michael and abby) simply cannot see how exploitative capitalism and white supremacy have any place in a faithful Christian life. Many Christians throughout the centuries have shared this conviction.

We come to this conversation, too, with a clear understanding that the fossil fuel industry, which has been undergirded by Christian theologies and communities, has made everyone complicit in climate change. (We'll say more about this relationship between Christianity and fossil fuels later in the book.) We are all collectively dependent on fossil fuels, which is why we need to change systems as much as we need to change ourselves.

We are not without hope, however. Whatever people do in rebuilding the economy of God, we must do it with beauty and with justice. And so, we write with humor and stories and sensationalism, unafraid to let the stories we tell hit you in the throat because in them you meet people who are transforming the systems as they are now. We tell our own stories too—stories of how we've organized as people of faith for climate justice, stories of how we have worked for just climate finance, stories of how we have learned from other leaders.

Two brief comments before we enter into the book together. First, a word about climate injustice. Climate injustice is rooted in the reality that the people least responsible for climate change are experiencing the first and worst effects of climate chaos. Climate justice seeks to right that imbalance.

Second, a word about method for this book and how you, as the reader, might engage. We come from different and overlapping worlds. Some of what follows reads as traditionally academic. Some reads as a sermon or a colloquial account. Our goal is to author this book as our whole selves, engaging different concepts and weaving together many different voices. And, as we wrote this book, we tried to be transparent about the dialogue that happens when two people collaborate. You'll see this dialogue as italics throughout the chapters. Sometimes we agree. Sometimes we disagree. Sometimes we push each other to think beyond what we concluded on our own. As you engage this book—its reflections, calls to actions, blessings, and dialogues—how will you add your story and the stories of your communities?

So, thank you for joining us in this book about moral economies, climate justice, and climate hope. May you be challenged, connected, and spurred to action, together.

Rev. Michael and rev. abby, PhD

CHAPTER ONE

Who Are We and Who Are You?

Introduction

Capitalism That Drives Climate Injustice

We first met in preparation for the New York City–based climate strikes in 2019. At the time, we were both working for nonprofits that supported and organized people of faith around caring for the earth. The youth organizers for the strikes wanted there to be broad faith representation at the march for the strikes, and our organizations stepped in and up to gather people for the faith bloc. We organized together, with spreadsheets and emails and webinars. We organized together, with prayers and rituals and presence. We organized together, as adult allies and witnesses to the powerful action youth organizers were demanding from older generations.

Neither of us is a stranger to the climate crisis. We have both organized with frontline communities, and both of us have organized in many ways with people of faith to respond to climate change. When we first started talking about a book on moral economy and climate justice, we started with how we understand God calling us to build community as Christians. In doing so, we looked to the early church as described in the New Testament and to the ancestors of our faith in the Hebrew Bible. The stories of the early church are examples of mutual aid and care, giving up what is held individually for the good of all, and living out the commandment to love one another. These stories undergird the book in your hands. In our minds, a moral economy needs to be built upon the root of the Greek word for economy—*oikos*—the ecology, the community of connections. Whatever money exists for a community, especially in a time of climate injustice, would exist for the community as a whole.

While some argue that the purpose of money is to make more money, others argue that the purpose of money also lies in meaning and identity-making. Christian biblical traditions affirm this in the Gospel of Matthew, which says, "For where your treasure is, there your heart will be also."[1] In this framework, the placement of investments and spending shows what issues and concerns are important to those who decide where and how they invest and spend. Economists and activists alike refer to this as a moral economy, "in which citizens use the power of their purse to inject ethics, reciprocity, and care" into economic life.[2] For us, in our contexts, we see how spending our money has often ignored biblical boundaries for creating the good life and increased the suffering of many through intersecting systems of oppression like racism, classism, and sexism.

Each of us—no matter who we are—will come to this work to build a new moral economy from our contexts. Honoring contexts is a cornerstone of the method of this book as we wrestle with economy, climate injustice, morality, and faith. We write this book in the same vein as Veronica Kyle and Laurel Kearns's "The Bitter and the Sweet of Nature: Weaving a Tapestry of Migration Stories" in *Grassroots to Global*. Kyle and Kearns share their own stories, occasionally writing in the first person. Kyle is a Black ecowomanist organizer who has spent most of her working life in Chicago. Kearns is a white sociologist of religion from Florida and an established academic in New Jersey. Their chapter honors the differences in their lived experiences, moving into first person when differentiation between authors strengthens the work or centers marginalized communities.

In the same way, we will write from our contexts. Rev. Michael Malcom is a Black cisgender male born and raised in Atlanta, Georgia. His pronouns are he/him. He is an ordained minister in the United Church of Christ. He holds an MDiv from the Interdenominational Theological Center and an MBA from the University of Georgia Terry College of Business. rev. dr. abby mohaupt is a white cisgender woman from northern Illinois, ordained in the Presbyterian Church (USA) and trained for a PhD in sociology of religion and social movements under Laurel Kearns. She uses she/her pronouns. We are preachers with backgrounds in organizing, teaching, and administration. While we have worked together in several contexts, we know that we are different people, rooted in a faith that demands that we work for justice for all if we will be faithful to God's call to us to love God, planet, and people.

What we know (and what we will explore in later chapters) is that the dominant story of economics conflicts with the story of the beloved communities that the biblical witness calls us to create. We've seen and experienced firsthand how capitalist mindsets have looked at theories and ideas about the beloved community and told us that they are wrong or naive.[3] We've seen and experienced firsthand how white supremacy has tried to make community and climate justice for only some, when all people deserve a healthy planet. In the following chapters, we aim to show how biblical traditions (specifically the story of Boaz and Naomi and the stories of the early church as told in the book of Acts) call us to justly practice caring for the community in all aspects of life, including our economic practices and policies. This call entails seeing community in the most expansive way possible. And from our vantage points, we have to see community as with God, each other, and the planet; otherwise, it's not sustainable and it is not just.

The core of our argument is twofold. First, if we are going to be faithful Christians, we have to reject fossil-fueled capitalism, as it contributes monumentally to climate injustice. This book will focus on North American contributions to the creation of climate injustice and to the strengthening of hope in the face of climate change, while also exploring the global implications of those contributions. Contributions to climate change are inextricably linked to capitalism. One major contributor to climate change is the production and burning of fossil fuels. This burning creates more emissions but also more profits for fossil fuel companies. In capitalism, there is no incentive to stop doing something that makes money even if it's devastating the planet and everything on it. Instead, our current form of economy says, "Make even more money, as much as you can, before the planet is all used up." It argues that this is our right as individuals—the right to thrive on our own, even at the expense of other people and our home, the earth. But the right to thrive hasn't been given to all people equally. In the following pages, we will explore how capitalism intersects with systems of oppression (particularly white supremacy) and supports the false narrative that as individuals, all of us can make it if we just try. To the contrary, because of systemic oppression, it is simply not true that everyone can thrive in capitalism. It privileges some groups at the expense of others.

Second, if we are going to be faithful Christians, we have to embrace God's economy of goodness for all parts of the planet. As a result of individualistic thinking and an immoral capitalistic economy, the planet is warming at alarming rates and people are suffering. If we are going to be faithful Christians, if we are going to survive as a species and a flourishing creation, we will not only reject exploitative capitalism and the systems of oppression that it affirms but will also build and practice God's economy. The current economy contributes to climate injustice in countless ways. Examples (primarily from the Gulf South because of the experiences we bring to the work) will be highlighted throughout the following pages.[4]

And just to be clear, *yes*, we are critiquing capitalism, but that does *not* mean that we are critiquing all business (we will even mention some of our favorite businesses in later chapters!). Instead, we are critiquing a capitalism that has become exploitative and predatory, capitalism that puts maximizing profit over the well-being of people and our habitat the earth, capitalism that enables some to become outrageously wealthy by exploiting others and the earth, capitalism that depends upon maximizing fossil fuel extraction and use, capitalism that allows megacorporations to wipe out small businesses, capitalism that was the foundation of the slave economy in this country.[5] By critiquing this form of capitalism, we are *supporting* the development of small-scale local or regional business, especially where it commits to a triple bottom line of people, planet, and reasonable profit. ("Triple bottom line" is a business framework that measures a company's success based not only on its financial profit but also its social and environmental impact.)

We are not hopeless. New (moral) economies that would allow for climate justice are possible, economies that align with the biblical call to care for one another and share abundance, a call that is relevant to our world here and now.[6] Real-life examples already in the making can guide us in the work ahead.

In the next section, we briefly introduce two ways we see moral economies emerging in the Bible. The stories of Ruth and of the early Christians in Acts reveal these two ways. These biblical stories will guide us through the rest of our explorations into new economies that are good for all. Thus, in the voice of a pastor, Rev. Michael, we invite you into a new economy for climate justice aligned with biblical tradition.

Check This Out

A New Economy Consistent with Biblical Tradition

First, I want to tell you that, if you are a Christian, you should read the Bible. I realize that many of us have been scolded into believing that the Bible is the equal to God. I am in no way advocating that harm. I am suggesting that you read your Bible because it is truly entertaining and sustaining. You will find comedy, romance, drama, and suspense in the pages of what many of us consider sacred text. The historical book of Ruth provides us with all of these elements.

First, a little more context about the book of Ruth. The text, one of the books in the Hebrew scriptures, part of the Bible, introduces us to Naomi, her husband Elimelek, and their sons Mahlon and Kilion. This family were Ephrathites. They were from Bethlehem, Judah. They moved away to live in Moab. While in Moab the sons married Moabite women, Orpah and Ruth. Elimelek died, a famine happened, and the sons died. Sounds like a number one hit on the country charts, does it not? Having lost everything, Naomi decides to return to her land. Orpah and Ruth want to go with Naomi, but Naomi attempts to discourage them. It works with Orpah; she falls back and returns to her family. Ruth cannot be dissuaded, however, so she and Naomi decide to make a life together.[7]

In the third scene of the book of Ruth, we are introduced to a man named Boaz. This history portrays Boaz as being wealthy, older, and noble. Naomi describes Boaz as their "close kin." This description of Boaz is that of a kinsman redeemer, someone who will take care of a relative in trouble. The kinsman redeemer had to be a blood relative, be able and willing to purchase forfeited inheritance, and be willing to marry the wife of the deceased relative. To build a moral economy it is helpful to understand the concept of the kinsman redeemer and how it applies to our care for creation. *(We're all interrelated—that is, kin—and many of us are in real trouble. —abby)* Boaz is not only named a kinsman redeemer; he acts out the role of kinsman redeemer in the life of Naomi and Ruth when he ultimately marries Ruth. A deeper look at the actions of Boaz provides for us an example of a moral economy.

When we are introduced to Boaz, he is surveying his land and his workers. As he approaches, he greets his workers. The salutation shows respect and connection. Boaz sees Ruth working in the field and asks about her. He knows of her but he does not know her—they are distant relatives. This is evidenced in the conversation that he has when he meets Ruth. Boaz knows that Ruth is widowed, that she has left her family of origin to be with her chosen family, Naomi, and that she is considered a foreigner in his land. Last, he knows that she is not wealthy. In essence, Boaz knows Ruth's burdens. Boaz responds to this information with respect and empathy. What Boaz does not do is choose this as opportunity to prey on her pain.

Unlike capitalism, a moral economy sees and cares for all. A moral economy does not use the pain of others as an opportunity for progress. It does not abuse its authority. It sees all as vital for ecological sustainability. When Boaz approaches Ruth he speaks to her humanity by acknowledging her experience and honoring her contributions. Capitalism ignores the experience of those who cannot participate. At the same time, capitalism benefits from the contributions of those who are poorest or marginalized. The story of Naomi and Ruth can be seen in the history of every marginalized person. I know all too well what it feels like to live in a place that is not my place of origin. I know what it feels like to live among people who see me as a foreigner and ignore my experience and contributions. Unfortunately, the story of Ruth and Naomi is not uncommon. Boaz and his actions, though, were uncommon then, and are uncommon now.

Boaz did not abuse his authority. He respected others. He acknowledged Ruth, he honored Ruth, and he invited Ruth into his life. Ruth was from Moab, a foreigner. The issue is, it was against the covenant to marry a foreigner (see Ezra 9 and 10). I can only imagine you saying, the more things change, the more they stay the same. Boaz's society was closed and did not welcome outsiders. Boaz's invitation to Ruth was an expansion of his community. It meant that he not only saw Ruth, he also appreciated Ruth. Ruth had value. Boaz shows appreciation for Ruth's value by inviting her in. A moral economy sees the values of others and appreciates it. Capitalism sees the value of others and seeks to profit from it.

Highlighting the actions of Boaz does not afford me the opportunity to explore the actions of Naomi and Ruth here. However, I will tell you that they were who I was thinking about when I recommended that you read your Bible. I'll sum it up by saying that these sisters are something else. *(Essentially, Naomi makes a plan for Ruth to go to Boaz one night to ask him to take care of her. Naomi is active in creating her future! She is not powerless or voiceless! —abby)* I can also say that Ruth must have been pretty special, because right after that night Boaz was making moves to marry Ruth. Not only does he invite her into his life, he assumes responsibility for her well-being, and makes a way for her future (as was necessary in the context of that patriarchal society).

Boaz acting as a kinsman redeemer means that Boaz participates in Ruth's future but he does not benefit from it. This is to say, as kinsman redeemer, Boaz would have seeded a child for Ruth but—in his culture—it would have been seen as the child of his deceased relative. The property that he purchased would be the inheritance of the child also. If the child decided to walk away from Boaz, there would be no recompense. This leads me to believe that Boaz acted out of a sense of community covenant, rather than personal benefit.

The moral economy that Boaz participated in was a deep commitment to community covenant. A covenant that saw and cared for community with selfless compassion and empathy. It was one of appreciation and honor for the other. It welcomed and expanded accommodations for the least of these. Last, Boaz's moral economy secured those who were less secure.

We see this same model at the birth of the New Testament church.

In the book of Acts, we are invited to witness the acts of the Holy Spirit in the life of the early church. The dramatic retelling of the manifestation of the Holy Spirit comes with symbolism that helps us to understand the call and command of the early church. The Holy Spirit is accompanied by tongues of fire. These tongues are said to have rested on everyone that was assembled. The first thing to note is that the tongues were evenly distributed. Everyone received the same portion. The other thing to note is that everyone received. No one was left out.

Understanding that no one was left out in the first action of the Holy Spirit at its manifestation follows the precedent and model of the kinsman redeemer in the story of Boaz, Ruth, and Naomi. We see that precedent carried forth by the early church. The manifestation of the Holy Spirit came at a time when the New Testament church was still in a state of confusion and fear. It was vulnerable. The Holy Spirit comes on the scene to rescue these frightened, confused, and vulnerable people to provide security, order, and mobility. Sound familiar? It should. This is the same model of justice and equity that we see in the kinsman redeemer motif.

After the manifestation of the Holy Spirit at the Pentecost festival, the early church begins life. When we meet the early church again, we find a community of believers has formed. This community is described as a sharing community who ensured the care of one another. The book of Acts said that they were selling their possessions to meet the needs of their community.

What is amazing about this narrative is that at a time when it would have been easier to return to life as usual, to maintain the status quo, the early church chose radical care and inclusion. Today, it appears that we are fighting to maintain the status quo and digressing as a result.

The text tells us that they sold what they had so that no one was without. I love that! Can you imagine a world where the needs of others were enough to make one want to share their wealth for the benefit of others? If you live in these yet–to–be–united states of America, you would probably have a hard time imagining that one.

Also amazing in this text is that this was a necessary step in the life of the church. The author of Acts thought it important to include the fact that they shared with one another so that none would be without. The gospel of inclusion was preached by Peter at the manifestation of the Holy Spirit when he spoke of the Spirit being poured out on all of us. Peter says that sons and daughters will prophesy in God's name. No one is left out.

It's radical care then and now because of the systems we operate in. In the times of the New Testament church, they operated in a caste system. There was very little chance of you rising above your born status. You were limited pre-birth. And it is the same for so many now. The capitalist system that we live in includes an economic caste system that is steeped in anti-Black racism. During the summer of 2023, I was led to do an environmental justice tour of the Gulf South states. We started in Florida's Panama City and ended in Miami. I met and spoke with residents in all five Gulf South states. I believe that a few of them lend testimony to what we have seen in the Ruth, Boaz, and Naomi narrative, and they are living into the model of the New Testament church.

I met with people like the Carroll family, who direct Green the Church Louisiana. Rev. Bruce Carroll is a Baptist preacher in Shreveport, Louisiana. The part of Shreveport where Rev. Bruce resides is rural and underdeveloped. The themes of abandonment, loss, and lack come up quickly when speaking to Rev. Carroll about the condition of his community. He describes development that has taken place without any impact assessments, causing flooding because there's no drainage infrastructure.

I spent time with my dear sister Roishetta Ozane of the Vessel Project Louisiana and her community in Lake Charles, Louisiana. They testified about the fossil fuel-burning power plants that surround their community and the ill effects on their health and community. They spoke of their land eroding because of human-made channels for runoff that pollute their water and land.

These incidents aren't isolated. Human-made unnatural disasters occur in Louisiana and Florida in the same manner. According to Lawanna Gelzer of Orlando, Florida, flooding was not an issue until they put an industrial park on wetlands and cut down the trees. Now, the flooding affects the more resourced communities just as it does the less resourced communities. The devastation from these human-made unnatural climate disasters affect them all.

It is greed and othering that drives decisions that allow a superfund site next to a community garden. It is greed and othering that leave collapsed infrastructure, abandoned homes, and toxic pollution in Black communities such as the ones we see in the northern parts of Birmingham, Alabama. The greed corrupts and the othering provides an excuse. This is what has created the toxic conditions in Gloster, Mississippi, as they battle against the biomass industry. The biomass industry is a prime example of efforts to combat climate change developed *without* considering their damaging impact on marginalized communities. This business has caused people breathing issues, skin irritation, and other health-related issues. Because Mississippi is paying this industry to be there, the community is left paying for their own poison. The community bears all the burdens with no benefits.

The othering of these communities is evident in the ignoring of their plight and fight. Many times they are left to fight on their own until someone sees how they can also benefit from the suffering of these communities. Big nonprofits and other agencies will swoop in for a photo op. Or they will come to speak on the issue for the cameras in the name of raising awareness, but they and their money leave when the cameras leave. Othering results in communities being abandoned and fending for themselves.

I was at the airport today while traveling to do ministry in Dakar, Senegal. I ran into a friend and fellow pastor in the nonprofit sector of ministry. As we were talking, we both expressed our anger with the church. We both came to the conclusion that it should not be this hard to do good. It should not be this hard to do what Jesus taught and the Holy Spirit modeled. Beloveds, I must ask, what are we doing? Where did we go wrong? What happened?

Today, our churches have replaced radical care with capitalism. We buy land, build buildings, hold events, and collect money. Where did we go wrong? What are we doing? But first, who do we believe we can be? In this book we explore that very question.

A Map for the Rest of the Book

This book is part of a larger series of books about building a new moral economy in the context of God's vision for the world. In the introduction to the series, Cynthia Moe-Lobeda casts a vision of ten "fingers on the hands of healing change" (ten forms of action).[8] Those ten fingers are:

Build the New—creating or supporting new financial systems, businesses, and agriculture that value people and planet over blind profit.

Change the Rules—changing public policy and laws to make people and governments more accountable and shrink the gap between rich and poor.

Move the Money—moving from a single bottom line to a triple bottom line, one that values people, planet, and profit.

Grow the Bigger We—building networks, alliances, and other forms of collaboration to heal exploitative economies in favor of life-giving and inclusive economies.

Resist the Wrong—saying no to extractive and exploitative economies.
Change the Story—recognizing the meta-narratives of economy and seeking a new story of what can be.
Live Lightly—changing individual action and lifestyle as a way to influence systemic action or personal values.
Listen and Amplify—paying attention to the stories of people who are on the frontlines and make those stories heard.
Practice Gratitude, Lament, Holy Anger—allowing ourselves to feel all emotions about the world that exists and the ongoing struggle for a better world.
Drink the Spirit's Courage—digging into the rituals and resources of our faiths in order to make a more just and equitable world. We'll be weaving this finger on the hands of action into the ends of each chapter, with blessings and actions for you.

These ten means of action, along with biblical economies, guide us in the work ahead. We cannot afford (pun intended) to *not* take up this work. In the following pages, we'll walk you through eight of these ten fingers on the hands of healing change (and more briefly engage the last two) as they intersect with a moral economy in a time of climate change. For us, that requires an honest grappling with the sins of white supremacy and capitalism. Each chapter will include how we see each finger working for healing, stories from people on the ground, an invitation to you to take action for climate justice, and a blessing for our shared work. We cannot make the world a more just place for all until and unless we do this work together, with faith.

CHAPTER TWO

Build the New

IN THIS CHAPTER, we're exploring the first finger on the hands of taking action for healing change in our economy: *Build the New*. In the context of climate injustice, this means we need to grapple honestly with how the old systems have contributed to environmental racism and how unmitigated economic growth has created chaos and suffering. We have to be willing to face the reality directly and clearly if we're going to create a new economic system based in climate-friendly and racially equitable small-scale business, community-based finance, and local/regional sustainable agriculture.[1] We need to be rooted in this new economic system if we take the Bible seriously. How else can we live into a vocation of taking care of the garden (Genesis 2:4b—15)?

In this chapter we explore the realities of climate change, the intersection of climate change and racism, and the role of Christianity in supporting both (and we'll do a little grappling with sin while we're at it!). Then we share examples of small-scale business, community-based finance, and local and regional farming in an economic system guided by sustainable values and the flourishing of our shared planet. We use a dialogical method, by which we mean that we have added our voices to this piece in ways that are obvious. We hope our conversation invites you to think and to take action. Throughout, we return again and again to our argument that if we are going to be faithful Christians in a time of climate injustice, we have to reject a predatory economy and white supremacy in favor of an economy that values the flourishing of all parts of the planet. That requires we build something new, but that newness itself requires that we understand just why the current system doesn't work—not for most people and especially not for people who have been marginalized.

Climate Change and Environmental Racism

For years I (abby) have been reading and reviewing the Intergovernmental Panel on Climate Change (IPCC) Assessment Reports. In 2007, the fourth IPCC Assessment Report began to persuade people that climate change is real, is caused by [some] humans, and is unprecedented. Before the report, there was not widespread belief in a changing climate, nor in a common human-made cause, though writers and scientists like Bill McKibben and James Hansen had been making those connections since the 1980s. Soon after that report, the United Nations Framework Convention on Climate Change said that humanity has "common but differentiated responsibilities" in response to climate change.[2]

I was struck by the language of *differentiated*. Why not assume that we will all take action on climate change and that our roles in that common fight will be similar? At the crux of a response is the reality that "global climate change exacerbates nearly all existing inequality" as a crisis that "dangerously intersects race and class."[3] So we earthlings face a catastrophe that is made more difficult by the shards of community broken by categories of difference and power. The world itself is broken, and we are broken too.

Still, perhaps not all hope is lost. I am struck by William Connolly's sentiment in his book *Facing the Planetary* that "the challenges today solicit both an embrace of this unruly world and pursuit of new political assemblages to counter its dangers."[4] Such an image is a reminder that when we recognize the real chaos of climate change, we do not respond to the breakdown of creation on our own, nor do we experience the effects of climate change in the same ways across categories of difference.

We are insufficiently prepared to respond to climate change together as a species. Truly, "we don't habitually place the well-being of others in the same framework as our own. . . . One 'race'—the human race—subjugates the rest of the biosphere for human benefit alone."[5] Our disconnection runs deeper than that—our species is racked by power imbalance and domination. Part of the human species has more access to global power and policymaking, and those parts of our species have historically been white. Certainly, we humans divide ourselves by other categories of difference: class, gender, sexual orientation, and physical ability are just some of the ways we cut ourselves off from each other. In this section of the chapter, I will home in on racism as a system of oppression in relationship to climate change, especially because of how Michael and I have seen white supremacy make climate change worse for communities of color.

I suggest that at its heart climate change is inherently racist, and thus a theological problem with which to grapple.[6] *(I would suggest that the climate crisis is exacerbated by what Rev. Dr. Benjamin Chavis coined "environmental racism." —Michael)* We will do this grappling in pieces that on their own reveal part of the larger story of racism in climate change. I'm writing this section with a deep sense of how the labor of dismantling racism is so often given to people of color, but it is the work of white folks. We who are white have to take responsibility. We also have to keep listening to our colleagues, comrades, and friends who are Black, Indigenous, Asian, and Latino. So, Michael breaks in here in places, not as a spokesperson for Black communities but as his own powerful voice connected to many communities.

Race and the Environmental Crisis

The first piece of the entangled problem is the connection between racism and environment degradation, which is well-documented. In his article "Whose

Earth Is It Anyway?" the late Black liberation theologian James Cone writes that global white supremacy has been deadly for Black people and Indigenous folks over the last five hundred years, and it has led to the exploitation of nature.[7] That is to say, the beliefs that people who are white have the right to control people of color are the same beliefs that say nonhuman species must be controlled. Ethicist Carol White expands on this connection by writing that the racism of the Enlightenment divided creation into the human and nonhuman binary, placed people of color in the nonhuman category, and then subjugated every being in the nonhuman category.[8] Reclaiming and acknowledging the humanity of people of color and the right of creation to flourish is key to unpacking environmental racism (and really the least that white folks can do).

In 1982, Black women and men in Warren County, North Carolina, lay down in the road to stop the incoming shipments of toxic soil to be dumped in their community. While it was not a successful campaign (in that the toxic soil still came into their community to be dumped), over four hundred people were arrested and it was the first time people in the United States were detained while trying to stop a toxic waste dump.[9] Subsequent grassroots campaigns against toxic dump sites have been occasionally successful, owing to the power of individuals coming together. As a consequence of that original campaign, the phrase "environmental racism" was coined in 1987 by Benjamin Chavis to encapsulate how pollution-producing facilities are placed in "poor communities of color," and how those communities lack important resources to access information and afford lawyers.[10] The connections between racism and environmental degradation should come as no surprise in a world that is manipulated by systems of oppression.

(Dr. Robert Bullard says that pollution follows the path of least resistance. As we explore the South juxtaposed to its history, we find that industrial pollution followed the path of redlining. Redlining was the time that white supremacy dictated where Black people were able to reside. They gave my ancestors the least desired land. I find it awfully funny that it was not desirous land for white people to live on, however it was great land for pollution. Fast forward to today, when Black communities are being gentrified and priced out of their generational homes. I guess that—to the white folks who are ousting the Black people from their communities—the land is okay now. Racism is giving me irrational![11] *—Michael)*

So irrational! As the protests in Warren County showed, there is an expectation that people of color would bear the economic and environmental cost of environmental degradation and that people who are white or of Western European descent would not. These expectations continue to arise through protests of "Not In My Back Yard" (NIMBY), disparity in public health, and immobility. Each of these issues arise because people of color continue to not have access, due to lack of political connections or financial resources, to policymaking or makers. Their voices—their humanity—have been systemically valued less than the voices and humanity of people who are white. While environmental racism emerges internationally, I am going to point to several examples just in the United States, for the sake of brevity. Later we will explore what a global pattern of environmental racism looks like in the problem of climate change.

When the 1987 *Report on Race and Toxic Waste in the United States* was published by the United Church of Christ's Commission for Racial Justice, it shed light on how race is the best indicator for where toxic waste is planned to be located.[12] Part of this reality comes from a NIMBY mentality—that reasonable people would not want pollution-creating sites in their neighborhoods. But because people of color do not regularly have access to places of power in governing bodies, it is easier to locate coal fire plants, incinerators, and oil refineries in communities of color. While this first gained public recognition in Warren County (as mentioned above), it has happened around the country. In Chicago alone, the Latino communities of Pilsen and Little Village have fought to have the coal plants in their communities closed, and the Black community in Altgeld Gardens has organized against the landfill surrounding their neighborhood. But NIMBY also indicates whose voices get heard when people of color do speak up. The Dakota Access Pipeline was originally planned to be built in Bismarck, which is a mostly white city. While the plan for the pipeline was eventually moved before the people of Bismarck could protest, one of the reasons was that the pipeline would threaten the water supply in the city.[13] The Lakota peoples asked for the pipeline to be moved for a variety of reasons, including risk to *their* water supply, but the pipeline stayed. The sacred camps at Standing Rock were resisting and were necessary because the state would not take no for an answer from the Lakota Standing Rock Sioux, nor was the risk of water contamination deemed too high. The mostly white city of Bismarck had a right to clear, uncontaminated water while the Indigenous people on the reservation did not. The people of Bismarck did not even have to protest, while the voices of the Indigenous people were ignored.

Access to clean land, air, and water is related to NIMBY, but it also is related to a public health disparity inflicted on people of color. People who are poor or are racial minorities bear a greater burden of cancer, asthma, and increased infant mortality due to environmental concerns.[14] These and other related health problems (like access to healthy food) disproportionately affect people of color. Two examples follow. The first one shows how the problem looks. The second one shows how grassroots communities are solving problems they face in their communities. First, the majority Black city of Flint, Michigan, was exposed to contaminated water that flowed through lead pipes. In 2014, the city's water source was switched to the Flint River, which had been the site of dumping from now-closed factories. When residents in Flint had to rely on bottled water for drinking (an environmental issue on its own), Representative Dan Kildee called race "the single greatest determinant of what happened in Flint."[15] The state's slow response to the water crisis and the effects of contamination led to health complications that would not have been expected in the rest of mostly white Michigan. Access to healthy food and clean water help mitigate health problems; without them a myriad of health problems emerge.[16]

Second, Richard Joyner and his predominantly Black community in Conetoe, North Carolina is one example of responding to the food desert and accompanying health problems like obesity and diabetes, which were literally killing Joyner's community. That community built a garden and an apiary so that they would have access to healthy food.[17] *(There are communities in the northern part of Birmingham, Alabama that are located close to a superfund site. They have been advised not to grow food in the yards because of the contamination. There are no fresh food sources located within the five miles of these communities. These communities have formed outside partnerships that help to provide fresh fruits and vegetables with pop-ups. One of the primary drivers of this initiative is led by a Black woman, Jilisa Milton. —Michael)*

(It's always the women, isn't it?)

Even if people can successfully claim their right to clean water, land, and living, sometimes health risks cannot be alleviated. In the past, when places have been exhausted or when food and water sources have become contaminated, people would be "forced to move, to look for another place pleasing to life, in order to survive."[18] Environmental racism emerges here in at least two ways—in the ability to stay in the places one calls home and in the resources to leave if necessary. These issues emerged in the wake of many natural disasters and were particularly acute after Hurricane Katrina. As the waters of the hurricane rose, people fled New Orleans if they could. But people in the Ninth Ward—the area of the city that was primarily Black—were unable to get out or had nowhere to go. They stayed in an area of the city that was poorly prepared for the devastation. The Ninth Ward still has not been rebuilt and the buildings that remain leak toxic fumes, whereas (white) neighborhoods with more access to political power have been rebuilt.[19]

(The Ninth Ward also brings to the forefront another truth. Disaster is equal, recovery is not. When the Ninth Ward attempted to rebuild after the storm, there were contractors that came in and took advantage of the victims. They took the insurance money of residents ill-affected and ran off with it. There are still blue tarps on homes from the Katrina disaster. —Michael)

Early conversations and movement building around environmental racism did not include climate change. Instead, they focused on urgent issues in local communities of color, like access to fresh food and clean air.[20] *(I am not so sure that this was a bad method. I think that we should go back to centering the immediate needs of the communities. Help the people, heal the planet. What affects the people is the same thing affecting the planet. The power plant in Memphis, Tennessee, is choking out that community and the cosmos. If you shut the plant down and switch to solar, the people are helped, the planet is healed, and advanced global capitalism is hindered. —Michael)*

YES! Those struggles are important and a matter of life and death, as we have seen above. If we know that racism infects all social structures and that environmental racism has been a historical and structural problem, how can climate change not be infected by racism? Climate change is environmental racism on a global scale.[21] In the following section I briefly document and illustrate the intersectional and interconnected problems of racism and climate change. Then we can begin to reconnect ourselves with each other to make a way forward. Truly, with climate change, the "we" must become a planetarily inclusive subject, or none of us may make it. That's what we mean when we say that we have to embrace God's economy of goodness for all parts of the planet.

The Problem of Race and Climate Change

So let's return to the definition of environmental racism: People who are white have created and enforced systems of power so that people of color live in areas that are targeted with environmental degradation. Those communities lack access to political connections and do not have access to information or the ability to broadly share their stories. As the implications of climate change become clearer over time, it is evident that "racism has denied the humanity of Black people" so that "tentacles of white supremacy have stretched around the globe."[22] That is to say, the countries that are hurt first and worst from climate change are majority people of color. Climate change is environmental racism on a global scale. Indeed "the people most affected by climate change . . . will be living in the poorest and more disadvantaged parts of the world," communities that are in the Global South.[23] One might wonder if the global community would have reacted more quickly to climate change if the Global North had been affected by climate change first, in a similar way to how the response to the water crisis in Michigan was slower in Flint than in whiter areas of the state. In the following, I will scale out environmental racism to see how countries of color have had to bear the weight of climate change, have had less access to global political power, and have begun to receive and share information about their experiences.

First, climate change creates sacrifice zones, areas that will be destroyed or altered as temperatures and sea levels rise. Naomi Klein's description of those sacrifice zones includes isolated, poor places where "residents lack political power, usually having to do with some combination of race, language and class." These are places and people that have been written off by privileged people in North America and Europe as places that do not matter.[24] Klein does not invoke the term "environmental racism," but her definition of the inequity of climate change echoes Chavis's definition in terms of systemic lack of power and experience.

Similarly, the fourth IPCC report includes a description of the ecological and human geographies—those places that are sacrifice zones—which will be hardest hit with climate change:

> Some systems, sectors and regions are likely to be especially affected by climate change. The systems and sectors are some ecosystems (tundra, boreal forest, mountain, Mediterranean-type, mangroves, salt marshes, coral reefs and the sea-ice biome), low-lying coasts, water resources in some dry regions at mid-latitudes and in the dry topics and in areas dependent on snow and ice melt, agriculture in low-latitude regions, and human health in areas with low adaptive capacity. The regions are the Arctic, Africa, small islands and Asian and African megadeltas. Within other regions, even those with high incomes, some people, areas and activities can be particularly at risk.[25]

Almost universally, race and poverty are interconnected (one of the reasons that we're talking about racism and white supremacy in a book about moral economy!). But in these geographic areas, even wealth will not protect people in the sacrifice zones from experiencing climate change.

Second, these sacrifice zones have less access to global political power. One example of how less access to political power will shape a community of color's experience of climate change is how Indigenous people relegated to reservations in the United States are affected. Decrease in access to uncontaminated water by climate change will affect native people's access to potable water. Their ability to adapt will be hampered by "historical and contemporary government policies," on top of damage to reservations, loss of hunting grounds, and tribal poverty. Particularly in Alaska, as ice melts and the landscape shifts, climate change will also mean that tribes will have to relocate, without "governance mechanisms of funding to support them," which in turn will cause a decline in community and culture, further economic difficulties, and isolation.[26] Indigenous peoples in the United States will have their lives and cultures upended by climate change.

These upheavals are happening around the globe. Klein points to how communities in less industrialized countries are damaged by global warming and poverty. She notes that the easiest ways to overcome poverty requiring burning more fossil fuels. Nevertheless, those countries are being asked (due to mindsets that demand all nations to reduce emissions equally) to forgo that development in order to reduce greenhouse gas emissions, even though industrialized nations had full use of such energy sources. Some efforts have been made to respond to this inequity and create a solution. The Greenhouse Development Rights were created to "recognize the West's greater responsibility for climate change."[27] Such a treaty understands the lighter responsibility of countries outside of the developed West and begins to balance the political scales of global power. This and other economic endeavors, like the "Marshall Plan for the Earth," seek to address the injustice of how industrializing countries, "whose people have contributed little to the climate crisis," are being expected to give up future development in order to address climate change. The concept of climate debt capsulizes these inequities that are intimately connected with global economic and political power:

> The climate is changing as a result of two-hundred-odd years of accumulated emissions, that means that the countries that have been powering their economies with fossil fuels since the Industrial Revolution have done far more to cause temperatures to rise than those that just got in on the globalization game.[28]

Like the Indigenous peoples in the United States, the countries that have less economic power are developing nations that will continue to be devastated by climate change, by little fault of their own.

Third, countries and communities of color that are experiencing climate change have discovered the challenge of telling their stories and having those stories be heard. The last part of Klein's definition of the sacrifice zones is that the "people who lived in these condemned places know they had been written off."[29] They are invisible, with less access to sharing information widely about their situation and less access to policymaking. They have to make people in places with power—places that will not be sacrificed in climate change—understand that their lives and stories matter too. The Maldives—a tiny archipelagic country in the Pacific Ocean which will disappear as ocean levels rise—has created a variety of political stunts in order to capture the world's attention. Before the United Nations Climate Change Conference in 2009, the president of the Maldives, Mohamed Nasheed, held a cabinet meeting underwater, bringing attention to how sea level rise is already affecting his country.[30] Such creativity will need to be engaged in order to capture global attention. One final example of how the countries in sacrifice zones are ignored: There is currently very little scholarship specifically related to racism and climate change. Again, this is why Michael and I are trying to raise awareness of this reality.

The silence in this area indicates disconnections among human beings, which is one of the ways we grapple with the theological concept of sin. That problem is where I want to now turn, before finally exploring how we might reconnect ourselves to other people and creation.

Sin as Disconnection from God, Each Other, and Creation

Conversation about the sins of climate change has numerous entry points, not least of which is a discussion of the biblical commandments to love creation and each other. Here I want to focus on the concept of sin in the world and in all our relationships from the perspective of Reformed Christianity, and especially as a Presbyterian minister.[31] This is not to say this is the only way to try to make sense of sin in the world (and Michael will add his two cents too!), just that it is the framework for how I have wrestled with it.

In the "Confession of 1967," one of the guiding documents of the Presbyterian Church (USA), sin is described as how humans turn against all other parts of creation and God, and turn instead to other things. It "not only describes sinful people as those who turn against God, but also as those who turn against fellow men and women, becoming exploiters and despoilers of the world."[32] Sin in this case is a disconnection from God, other people, and creation. This last piece—disconnection from creation—is especially striking in a chapter that has so far been concerned with human social constructs and systems. *(Especially striking as we see society becoming more and more self-consumed. —Michael)* I am convicted by Carol White's argument that if we only see humans as important, we do not go far enough in liberation or in theology.[33] We cannot ignore that we are part of an ecosystem that requires diversity—and that that diversity includes the human and other species. (Humans would not exist for a moment, for example, without the thousands of microbial species that live in our bodies and in a square foot of fertile soil.)

(But, I would argue that we do not even see humans as important. We see self as important. Anything outside of the self is not prioritized. This is how we other and ignore struggling mothers and elders who can hardly pay for food while trying to keep their lights on. This is the reason that we can remain silent as children are bombed. —Michael)

Reformed theology describes the human beginning as one in which humanity is made in God's image. Our relationship with God stems from our original state of being created in the image of God and given life to live in dependence and obedience.[34]

(The definition of reformed theology is flawed according to Carol White. Carol White says that if we only see humans as important, we do not go far enough in liberation or in theology.[35] *Reformed theology describes human beginning as one in which humanity is made in God's image. The emphasis of this theology is on humanity. From an ecotheological perspective, this does not go far enough. Ecotheology considers the whole household of God. It makes real the scripture found in Psalm 24 that says, "The earth is the Lord's and all that is in it, the world, and those who live in it." This is saying more than that it belongs to God. It is saying that it is a part of God. I argue that all of God's creation is made in the image of God. I wonder what the world would look like if we all understood the image of God like this. —Michael)*

Yes. When we forget this image of God, we can forget how much we depend on God. When we act as if we do not depend upon each other—or if we exploit the world around us—we become convinced that we do not need God or the rest of creation.

You said something! Keep going!

This is what I mean: We belong to each other, meaning the whole of creation and God. When we forget that, we sin—against God, each other, and creation. And that's wrong on so many levels. Methodist pastor David Atkinson describes this as a sin of disconnection, a sin that happens when we give up "our covenantal place in relationship to God, to each other and to creation."[36] Presbyterian theologian Shirley Guthrie suggests that part of the breaking of the human relationship with God happens whenever those in power seek to be God by "dominating everything and everyone outside ourselves."[37] This act of domination is clearly done in climate change and racism.

(I would say that this act of domination is clearly done in race relations, which exacerbate climate change. Environmental racism in practice is the main driver of climate change. If those smoke stacks were not permitted to be in my community, they likely would be deemed unsafe and not allowed. My community has been othered; this is why there are warehouses within five miles of where I grew up.

And it's why I could grow up far away from warehouses, even though I was raised in a manufacturing town. I got to grow up within a family that had a lot more control over where we lived and what was constructed around us, simply because we are white. And this is true about whiteness and white supremacy. One group of humanity has controlled part of creation at the suffering of both creation and a particular part of humanity—that is, people of color. If disconnection from God is one way to think about sin, climate change and racism are prime examples of that sin.

This is true on a global scale. Society at large has turned to consumerism. The "developed" countries are more consumer driven. We do the most harm and pass on the suffering. We have turned away from God and God's creation to capitalism. Everything is commodified, and we attach value to what we can commodify. This is why we profit from the pain of others. We would rather celebrate their resilience than to act to relieve them of the burden. This is how we don't see the hypocrisy in seeing vacant homes in our communities while demonizing our houseless brothers and sisters. —Michael)

And so our relationship with each other is sinful when we become disconnected from one another. Guthrie contends that one of the ways we disobey God is through contempt for any human being.[38] Instead, Reformed theologians imagine humanity in relationships in mutual openness and help—relationships that we clearly distort when humanity does not see people of color as people. Daniel Migliore writes that humanity rejects "our need for our fellow creature, most particularly those who seem strange or 'other' to us" when we sin.[39] This description returns us to Klein's sacrifice zones—communities of people whom people with power see as the "other." The racism implicit in climate change is a sin of disconnection because it does not see the humanity of people around the world as also made in God's image.

But climate change also disrupts our theological relationship with creation. That relationship is embedded in our biblical beginnings. In Genesis, one of the stories of human creation is God breathing life into our topsoiled-formed humanity.[40] Karen Baker-Fletcher (who trades "topsoil" for "dust") sees the role of dust used in the biblical stories of creation as a "metaphor for our bodily and elemental connections to the earth—our dustiness refers to human connectedness with the rest of creation."[41] And biologically speaking, it is not just dust that connects us, but water—we are bodies of water within a body.[42] When we forget those spiritual and biological connections, all of creation suffers. Included in part of the definition of environmental racism is the fact that the voices of those who suffer from environmental degradation are silenced—and this is true for so much of creation. The groanings of creation have been cut off and disconnected from humanity.

All of this discussion of sin has been from the perspective of a white human in the world. But that is clearly not the only human context. The definition of sin will be different for people who are white, people who are Black, and people of other identities. For people who are white, the sin of disconnection is rooted in our contributions to white supremacy, and the act of erasing the humanity of people of color.

(I will not jump in to speak for all Black people. I will speak of my experience and my understanding of sin. However, I must start off by telling you that I grew up Pentecostal Holiness. No, not the Church of God in Christ. We weren't Apostolic either. We were this hybrid of those and Baptist. The worship was highly animated. For those of you who know Black church, you will understand. For those who grew up in other traditions, prepare to be entertained. I went to a family church. In fact, I was a part of a family denomination.

We would be in church services all week long and three times on Sunday. There was prayer service, Bible study, Sunday school, morning worship, and evening service. I eulogized my cousin's funeral who died suddenly. We grew up side by side because our families lived so close. If I was not at his house, he was at my house. He had long stopped going to church prior to his death. During the eulogy I told those gathered that we were made to go to church so much growing up that I didn't think God minds if he took some time off. I think he had plenty of PTO [paid time off) stored up.

The sermonic thrust was fire and brimstone. Sin was emphasized heavily. I mean everything was a sin. If you smoked or drank alcohol it was considered a sin. If you engaged in sex outside of marriage, sin. If you engaged in certain acts during sex while married, you sinned. If you wore short pants, sin. If you listened to secular music, sin. If you wore makeup, sin. Everything was a sin! In fact, my going to seminary was/is considered a sin. It was in seminary that I became more aware of the concept of sin. It was in the environmental justice space that I learned the true definition of sin.

At the root of sin is selfishness. It is when we become so self-centered that we become disconnected from God and creation. This is how you explain how we pass by communities experiencing environmental injustice to go to more livable communities while disparaging the communities that our consumerism is harming. A disconnection from God and creation explains why we are not flooding the offices of every state legislature and regulator demanding solar and putting an end to energy inequity. This disconnection is how we can see a Black man's life choked out of him, a woman dying in her cell, and another being shot to death, and, when Black folk respond with "Black Lives Matter" in care and defense of Black lives, others respond with "all lives matter." In these instances, we miss seeing God and God's creation.

Capitalism is the biggest sinner of all. We live in a capitalist society. The problem with capitalism is that it is driven by profit and only sees money. Capitalism does not have a soul. It does not care. It does not know kindness or compassion. It knows revenue and rewards greed. Our relationship with capitalism is perverse. Capitalism drives us! Environmental racism is driven by capitalism and ignored because of it.

In order to break our cycle of sin, we must break from our relationship with capitalism. Then, we must move forward together. —Michael)

Ways Forward in Assemblages

We can't really build the new until we understand the wrong of what was. This is important! And it is how we find a theological way out of the sin of environmental racism implicit in climate change—we reconnect and reassemble the relationships between us and other people and between us and creation (and then the economy). The climate will continue to change, even if we do nothing. What we *can* change is how we move forward responding to the racism implicit in climate change, and how we move forward to lessen the degree of climate change that will happen if we do nothing. We can do both by changing the systems that have led to the power imbalance and to massive climate change. Those systems include not only economic policies, practices, and institutions but also the storylines that shape our lives.

Whitney Bauman writes that "every good movement needs a narrative. These stories are the threads that tie our own lives into a meaningful" history. While the story of the environmental movement has privileged a white, Western, and male story, stories can be retold and reheard.[43] In the retelling we can reconnect the pieces of environmental racism, climate change, and sin into a new whole. It's why I'm not the only one telling the story while also trying to take on the heavy lifting of explaining racism to people.

This exploration of the racism of climate change is not a conversation just about me. Individuals acting alone are insufficient to grow a movement; groups of people acting together must mobilize in new systems. In order to do so, we who are white need to act in solidarity, need to listen to and take direction from people of color who are already in the work, or who long to be if we would just shut up. People of color are not voiceless in the fight for climate change, but we who are white often drown out their voices. We have to be willing to listen and to share power. In doing so, we participate in the humbling, reconnecting work of God.[44]

The solidarity of people of privilege[45] with people who have been systemically harmed by climate change should be on terms set by those marginalized peoples. This kind of solidarity makes space for an assemblage of responses to climate change and the racism inherent within it. It reconnects people, overcoming the disconnecting sin of racism. And it makes space for a gathering that is already being assembled through creativity, radical listening, resistance, action, and community-building that allows the marginalized to be seen and heard.

There are countless examples of creative responses to climate change—collectives of people who are telling stories, creating art, writing poetry, and singing songs to rebuild connections that climate racism breaks down, generating new climate policy and energy policy, building and supporting small local farms, including Black-owned and Indigenous-owned farms. One example is the Wisdom Keepers Delegation. This global delegation of Indigenous peoples is advancing contemporary climate policies by distilling the wisdom of Indigenous cultures. Their activists and traditional knowledge holders are dedicated to integrating Indigenous perspectives into global climate action. They come together to offer solutions rooted in harmony with nature, ensuring that Indigenous voices guide the path toward a just and sustainable future for all.[46]

But in order for stories to be effective, they need to be heard, particularly by people who have systemic power. (We dig more deeply into this in chapter 9.) Feminist theologian Nelle Morton called for people to practice hearing voices into speech.[47] Some voices that continue to be drowned out in the experience of global environmental racism in climate change are the voices from Standing Rock and the ongoing activists working against pipeline sitting on reservation lands across the country. They join with the voices of the participants in the Green Belt Movement in Kenya, who continue to plant trees to fight back against the notion that vulnerable communities are helpless. Their plantings are a different kind of listening—to the voices of earth, which cry out to be rejuvenated.[48]

The 2014 People's Climate March illustrated intersectional resistance, with many of the marches focused on environmental justice in the fight against climate change. Hundreds of thousands of people assembled in the streets around the globe—their bodies mimicking the bodies of those in Warren County in the 1980s—to raise their voices together against the systems of oppression and greed that have created climate change and racism.

Each of these pieces remind us that "perhaps it is possible to work toward a common destiny together," knowing that our destinies are entangled.[49] Individual action is important for changing our own hearts and getting our lives aligned with our faith, but the devastation of our planet has been a systemic process over the course of generations. Undoing it, or even mitigating it, is going to require communal reimagining and reassembling. When it comes to climate change, we have to think about community in a global sense. We must start to recognize that our communities are mutually dependent upon each other for salvation. And we cannot let the racism implicit in climate change and the world disconnect us anymore.

Community-Based Finance, Business, and Agriculture

So far, we have pointed toward what has been, including how white supremacy and climate change have been comingled. Here we introduce three forms of "building the new" at work in our world today to build climate justice: community-based and values-oriented finance, business, and agriculture.

Climate-Friendly and Socially Equitable Financial Systems

In later chapters, we'll explore how Christianity undergirded the fossil fuel industry's expansion and then how Christianity has been part of the movement to divestment from fossil fuels.[50] In organizing within that particular movement, we faced opponents to divestment who worried that creating something new (in this case, investment portfolios that didn't include fossil fuels) would create harm, namely loss of income, loss of jobs, and loss of space at the table for shareholder engagement. It was also argued that something outside of the traditional investment model simply couldn't be done. But examples abound of community-based financial systems that value the planet in an economic system that is guided by values and the flourishing of our shared planet. These systems embrace a triple bottom line, or what will support people, planet, and profit.[51] And they reject a process that strives for as much profit as possible regardless of the consequences. Such systems already exist within the finance industry and communities.[52]

While Christianity has been aligned with capitalism, many leading voices within the church also have been critical of it and of any economic system that prioritizes wealth accumulation for a few over human well-being. In her book *Toward a Better Worldliness*, theologian Tara Rowe writes that Christianity has "responsibility for a tradition entangled with both the rise of global capitalism and the modern environmental movement."[53] Yet, she notes, while Protestantism has had a leading role in constructing capitalism, it hasn't been without strong criticism from within the church. In his book *Alternatives to Global Capitalism*, Ulrich Duchrow explores how both Luther and Calvin pushed back against early capitalism. Luther in particular called on priests to withhold communion from capitalists who practiced extractive usury.[54]

(The relationship with Protestantism and capitalism has deep roots. According to Kevin M. Kruse, who wrote One Nation Under God: How Corporate America Invented Christian America, *American Christianity was built and paid for by corporate America. In his book he says, "The postwar revolution in America's religious identity had its roots not in the foreign policy panic of the 1950s but rather in the domestic politics of the 1930s and early 1940s. Decades before Eisenhower's inaugural prayers, corporate titans enlisted conservative clergymen in an effort to promote new political arguments embodied in the phrase 'freedom under God.'" He goes on to add: "By the late 1940s and early 1950s, this ideology had won converts including religious leaders such as Billy Graham and Abraham Vereide and conservative icons ranging from former president Herbert Hoover to future president Ronald Reagan. The new conflation of faith, freedom, and free enterprise then moved to center stage in the 1950s under Eisenhower's watch."*

The sad historical reality is that our American Christianity was corporate-shaped and is—in part—corporate-owned or financed. This perversion is linked closely with Christian libertarianism. When you read, "In God We Trust" on your money, think about the God that it represents.[55] *—Michael)*

We need to reclaim the God of beloved community. We will explore how this community emerged in the early Christian church, as depicted in the book of Acts in a later chapter. Here we glimpse it emerging in community-based, values-oriented finance.

In 2004, Peter Krull envisioned a financial investment firm that would allow investors to make financial choices based on their values. In 2010, Krull & Company's responsible investment strategy was selected as a case study for the United Nations's annual report, "Principles for Responsible Investment."[56] Soon after Krull became an informal adviser to many in the divestment from fossil fuels movement, especially as a voice to counter those who said that fossil-free investment was impossible. I, abby, first met Krull on a divestment from fossil fuels panel, where we both suggested that fossil fuel divestment was a sustainable and profitable model, and I added that this was a faithful way forward.[57] Krull and Company continue to be on the forefront of fossil-free investments. On their website, they describe this kind of reality in the category of divestment and reinvestment, saying that:

> Divesting means to remove something and place it (reinvest) in something else. Earth Equity Advisors offers clients the opportunity to invest without owning fossil fuel companies.
>
> Over the past several years, there has been a major push to divest from fossil fuel stocks—coal, oil and gas exploration, excavation and transmission.

Why would an investor want to divest from fossil fuels?

- To align their investments with their values
- To have a voice in public policy and limit influence by well-funded energy companies
- To potentially reduce their financial risk from energy price fluctuations
- To reallocate to more sustainable companies[58]

This is just one small illustration of a burgeoning movement to create finance that honors the well-being of earth, its climate, and all people.

Climate-Friendly and Socially Equitable Businesses

The movement to create and support forms of business that are committed to economic equity has a long history in the US and other parts of the world. It has expanded in recent years to include more fully the aims of racial equity and ecological well-being. Currently this movement is burgeoning with energy and innovation. At its forefront are worker-owned businesses ranging from tiny local cooperatives to the large-scale Mondragon Corporation with nearly a hundred thousand employees. Even cities—such as Cleveland with its employee-owned Evergreen Laundry, supported by the city's network of hospitals—are experimenting with building mutual aid and stronger communities through employee-ownership of business. The Worker-Driven Social Responsibility Network and its programs like the Fair Food Program enable consumers to support businesses that prove their adherence to human rights standards for workers. Many companies are daring to vastly reduce carbon emissions with the aim of reaching zero emissions. In this section, we offer two brief examples of businesses that adhere to this triple bottom line of people, planet, and profit rather than maximizing profit regardless of the cost to people and planet. We invite you to explore multiple examples and envision the possibilities emerging.[59]

Arrayed is an ethical clothing brand rooted in ecological and relational harmony. The company's founder, Liz Case, was intent on finding ways to create and sell clothing that wouldn't harm workers and would affirm her Christian commitments. She writes that,

> Arrayed is a liberative people-first, planet-focused Christian apparel brand that aims to align the call for justice and wholistic flourishing we find in Jesus's life with the messages on and production of Christian t-shirts. The Christian apparel industry is a $4.5 billion industry annually, yet many of the current options don't offer progressive theological messaging or transparency about their production process.[60]

The clothing sold through Arrayed is made by the Telastory Collective in the Philippines and WORK+SHELTER in India. Workers are paid living wages and materials are sustainably sourced. Each piece of clothing carries a message of progressive theology or teaching, with images designed in collaboration with other ministries. Indigenous and queer artists and pastors have partnered with Arrayed. While Case sometimes travels for in-person pop-up stores, everything is available online.

Another example, Koinonia Farm, is a community rooted in a new economy that rejects white supremacy, embraces sustainability in the context of a global reality, and seeks to embody biblical values. They describe their beginnings as,

> a "demonstration plot for the kingdom of God." This meant a community of believers sharing their lives and resources, following the example of the first Christian communities as described in the Acts of the Apostles. Other families soon joined, and visitors to the farm were invited "to serve a period of apprenticeship in developing community life on the teachings and principles of Jesus."[61]

These values were not always well-received by their larger community. As they say,

> Our commitment to racial equality, pacifism, and economic sharing brought bullets, bombs, and a boycott in the 1950s as the Ku Klux Klan and others attempted to force us out. We responded with prayer, nonviolent resistance, and a renewed commitment to live the Gospel.

They also responded by starting a bakery and mail-order business that would reflect the community's values. The store's website describes this commitment:

> By using fair trade, organic, natural, and sustainably produced ingredients, even our baked goods support our vision statement of love through service to others. We are always finding new ways to be sure our food remains delicious, good for the land, and good for those who grow the food and ingredients that go into our food.[62]

Koinonia Farm and Arrayed demonstrate how communities of faith and their allies are building new forms of business that honor both social equity and ecological well-being. Where do you see such businesses in your part of the world? How could you support them? What joy might you find in doing so?

Climate-Friendly and Socially Equitable Agriculture

Sister Grove Farm in Van Alstyne, Texas, is embracing climate action, ecology, local economies, and faith. This regenerative farm is run by Sarah (a minister in the Alliance of Baptists) and Rodney Macias, inspired by the Genesis teaching that God took the human and placed the human in a garden to farm and care for it (Genesis 2:15). Regenerative farming is based on the idea that we can heal the soil by tending to it, and by doing so we can be healed ourselves as part of the larger whole of creation. Sarah and Rodney have tended this particular farm that includes chickens, cows, sheep, and cover crops since 2016. They write that while they sell a variety of products (and host events), "at Sister Grove Farm, we do not see food as a product. Nor do we consider it to be fuel. Food is all about relationships."[63] This interplay of relationships can effect who we are in the world, as people of God connected to other parts of the planet. On their webpage about how to buy their chickens, they try to describe how regenerative farming works, saying, "Our chickens are foragers. They enjoy a diverse diet from what the pasture offers them. In addition, we supplement with a locally sourced non-soy, non-GMO feed."[64] These chickens forage in rotation with cows and sheep, in harmony and in a pattern that allows for the pastures to be nurtured by the waste and movement of the animals.

The farm is a business, one that invites the buyer into a way of interacting with the food economies, especially in response to industrial agriculture:

> The small act of your buying a heritage breed chicken might be your own personal resistance against an industrialization of our food system. It could also just be considered good sense and a renewal of a practice lost not that long ago. We believe that once we experience the taste of real food and its effects on our health, we will be transformed and can then affect our small corner of the world in a meaningful way.[65]

Sister Grove Farm is just one example of farming in a new moral economy, just one way that farmers are building the new.[66] Other examples abound. We invite you to enjoy learning more about this movement to build equitable and ecological ways of growing food. Some windows into its many dimensions include: the Black Food Sovereignty Coalition, the US Food Sovereignty Alliance, *La Via Campesina*, the Central Virginia Agrarian Commons, the Black Church Food Security Network, and Agrarian Trust. Two shining examples of specific farms are Soul Fire Farm on the East Coast and Planting Justice on the West.

Indeed, many communities and groups already are building the new, based on a morality that will not be quietly complicit with climate change. (Really, go look up the mutual aid communities in your town!) In this chapter, we have shown how climate change and white supremacy have been enmeshed and how communities are starting to build new forms of finance, business, and agriculture—lynchpins of the economy. Throughout, we've tried to show how Christian faith invites us into a different way forward. Are you ready?

Invitation for Action

Michael: I sat in Riverside Baptist for a meeting with the National Council of Churches in 2019 during New York Climate Week. I heard clergy from all over the world celebrating the announcement that churches and other faith-based institutions had divested over $100 million from fossil fuels. I was moved to inquire about the outcome. What did you do with the money after you divested? There was no good answer to my question at the moment. I told those attending the summit that it is not enough to celebrate divestment. The church should be taking the lead on reinvesting finances to make a positive impact, especially investing in efforts that counter climate change and racism. What would it look like for the church to lead on the solutions? What would it look like for the church to offer practical implementation of paths toward climate justice to go with the thoughts and prayers? What would it look like for you to be an ambassador for care?

abby: And so you might be wondering what it would mean for you to build something new in terms of how your money impacts the climate and other people. Maybe you wonder if you could somehow make your treasure go where your heart is, as the Gospels suggest. Maybe you are asking yourself, "Do I need a better bank?" And maybe you're wondering if it's even possible to get away from the biggest banks, like Bank of America, Citi, Wells Fargo, and Chase, which invest heavily in fossil fuels. It is!

Our friends at Green America, whose mission is to "harness economic power—the strength of consumers, investors, businesses, and the marketplace—to create a socially just and environmentally sustainable society,"[67] has created a database of banks, credit unions, and other financial institutions. You can apply many different filters to your search to find a bank that meets a variety of criteria that are important to you and your values. For example, when abby was getting a new bank, she wanted one that would be smaller, invested in sustainable solutions, and had a history of racial justice. The bank she landed on was founded by immigrants and takes climate risk seriously.

You can make your own changes in where you bank by using their filters here: https://greenamerica.org/get-a-better-bank. A *next* step could be motivating your congregation, denomination, or place of work to divest from banks that fund the fossil industry and reinvest in finance institutions that support a healthy climate and social justice. From there, who knows? You might organize to get your city to join other cities around the world who are divesting and reinvesting.[68] Chapter 4 explores divestment from fossil fuels on an institutional and investment level and invites you into that work. But here, in this invitation to action: What will you do with your day-to-day banking so that it honors your values and supports what you believe to be good for earth and its people?

To learn more and to meet a global community of companions, be sure to visit the website for this book series at https://www.

buildingamoraleconomy.org/. Select "Book 1: Climate." There you will find a treasure trove of resources—organizations and websites, reading material, film, and more—to accompany you on the faithful journey toward climate justice.

Blessing

This is a blessing for your body. A blessing for the body you inhabit.
Put one hand on your chest and one hand on your belly.
Feel your feet on the floor.

Breathe in. Breathe out.
The breath that is moving through your body is connected to the breath of all other living things.

Breathe in. Breathe out.
The breath that is moving through your body connects you to the planet.

Breathe in. Breathe out.
The breath that is moving through your body connects you to other bodies on this planet, bodies that deserve clean air and clean water and clear soil no matter the body.

Breathe in. Breathe out.
The breath that is moving through your body connects you to the planet.

Breathe in. Breathe out.
The breath that is moving through your body connects you to other bodies on the planet, bodies that are strong and brave and need to move and rest no matter the body.

Breathe in. Breathe out.
The breath that is moving through your body connects you to the planet.

Breathe in. Breathe out.
The breath that is moving through your body connects you to other bodies on the planet, bodies that matter. If you are a white body, how can you breathe so that others may breathe too? If you are a body that has been oppressed, how can you breathe so that you may find rest?

Breathe in. Breathe out.
The breath that is moving through your body connects you to the planet.

CHAPTER THREE

Change the Rules

YOU ALREADY KNOW how bad climate change is, and you know how deeply connected it is to white supremacy and capitalism. These destructive systems are why we need to build something new together. And in order to create a different world that isn't wrecked by climate change, white supremacy, and capitalism, we have to change the rules that have shaped the world as we've known it as we live into a new community (a new *oikos*, the Greek root word for economy—and ecology!).

I (Michael) recently saw a clip of a pastor who remarked that giving to the poor only helps the poor. He went on to say that paying your tithes is what blesses you, implying that your giving should really benefit you. He juxtaposed his thoughts to the law of multiplication. I normally won't comment from a clip because a clip does not give context. However, this thought is a complete thought and not well-informed. *(It's also a very common thought. —abby)* His thoughts promote a false narrative that is based on a prosperity gospel that is deeply seated in some parts of the church, which suggests your giving should benefit you. It is perverse and ignorant.

There are far too many problematic examples of how the prosperity gospel exploits what hospitality looks like in God's view. These examples show God's blessings for the giver as a quid pro quo rather than highlighting the grace and goodness of God. There is also a deeper false narrative that I would like to address. When did the theology of the church become so self-centered and capitalistic? The pastor's emphasis on doing good to be rich is an issue. It says that we treat God and others like an ATM. You put in expecting something out. At the establishment of the New Testament church, the people sold what they had and took the proceeds to care for the community. In Acts 5:1–11, you find Ananias and Sapphira. When everyone sold all they had to care for the community, Ananias and Sapphira withheld. This is the one time you see someone looking out for their own self-interest. They were met by death. Perhaps society is so off-balance because the church is off-balance. Our sacred scripture teaches us how to live as a community. Providing for the needs of the poor is a mandate. Loving your neighbor is a mandate. It does not teach us how to get rich. Loving our neighbor and caring for our communities are part of what it means to be Christian, especially in a time of climate change.

To keep loving our neighbors as the climate changes will mean we will have to change the rules of what has been normalized in the past. In her book

Building a Moral Economy, Cynthia Moe-Lobeda writes that major implications of changing the rules of capitalism include "efforts to change public policy so that it reduces the gap between rich and poor; supports ecologically regenerative business, transportation, food systems, recreation, households, and more; and enables power to be distributed and accountable . . . [to] support policies that build equitable, ecological, and democratic economies."[1]

It's so hard for us to change; people do not like change (people also don't like the truth). In this chapter, we lift up how the rules we need aren't new at all—but rather are rooted in the early church. So, first we will explore how a significant part of the early church was set up. Then we will explore the Earth Charter, one example of how global rules are already being changed in ways that align with early church rules which set up systems of life together that provide for the poor and love the neighbor. Finally, we will illustrate how we might change the rules on local, national, and international levels in accord with the Earth Charter principles.

Early Church and Mutual Aid

The early church was set up in a different way than the empire that surrounded it. These early followers of Jesus were rooted in how he was calling for something different in the Kin-dom of God. Jesus was calling for a world that brings good news to the poor, release to captives, and sets the oppressed free (Luke 4:181–9). Elsewhere in the Gospels, Jesus redefines justice, replacing retribution with love and forgiveness (see Matthew 5:384–8). Jesus was killed because the Roman empire did not like this reordered reality, especially because it pushed against the economic dominance of Rome. This reordered reality demanded a moral economy, an economy rooted in care for other people. That's what happens when you push back on the powers that be. Jesus threatened the political and religious leaders of his time. Rev. Dr. Anna Blaedel, cofounder of *enfleshed*, writes of Good Friday as "empire's violence. . . . This is a day for bearing witness to all the lives and life destroyed through theologies of dominance, institutional sacrifices, imperial and colonizing forces, and stories that sabotage from the inside out. Lands and peoples, spirits and futures, creatures and communities. . . . We remain followers of defiant love."[2] It is dangerous to push against the norm. In the shadow of Jesus's murder, his followers scattered.

But then the early church emerged. The authors of the book of Acts record how the early church shared resources, living in a way that rejected the Roman empire's economy of putting wealth in the hands of a few: "All the believers were united and shared everything. They would sell pieces of property and possessions and distribute the proceeds to everyone who needed them. Every day, they met together in the temple and ate in their homes. They shared food with gladness and simplicity" (Acts 2:4–7, Common English Bible).

This wasn't a perfect community. Arguments arose and people were left out. The early church had to create a new system of leadership and new rules so that everyone would be fed. And so, deacons were appointed to ensure fair distribution of resources, especially to widows. Such a community decision highlights early structural change to support marginalized groups (see Acts 6:1–7).

Climate change requires new rules for us now too. "In the shadow of a hurting planet and hurting people, we need rules that do not glorify the wealthy. We need new laws and systems that value community and collaboration. It hasn't been easy to imagine new rules, and it's been even harder to enforce those rules. But new ways of thinking exist, new rules are possible."[3] And, indeed, if the old ways are rooted in white supremacy and exploitative capitalism, we need to abandon the old ways. New rules already are emerging, on local, national, and international levels.

Rules for Today

The Earth Charter in the Anthropocene

The Earth Charter has sat in the shadows of my (abby's) mind since I began exploring the intersections of religion and ecology in a time of climate change twenty years ago. The Earth Charter was begun by the United Nations, created by a process that included voices from around the world, and finalized in 2000. Like the Universal Declaration of Human Rights, the Earth Charter is a "soft law" document, meaning that it guides global morality and not international law.[4] This four-page document includes sixteen principles on respect and care for the community of life, ecological integrity, social and economic justice, and democracy, nonviolence, and peace.

The charter begins with a preamble that roots itself in the urgency of climate change and the need to respond in "a time when humanity must choose its future." It articulates that, as the earth becomes more fragile, the human community becomes more integrated. Humans share a common home and have responsibility to each other, the earth, and the future. The preamble uses the pronoun "we," indicating both that the writers stand united and that no part of humanity can escape the interconnectedness of life on earth. Earth is named as the home that is unique in the universe for supporting life, but is limited and needs to be protected. Here the writers include a nod to the sacred in humanity's responsibility to care for the earth and the life on it. The charter does not shy away from the multitude of ways suffering has impacted human and earth communities. Poverty, war, and environmental degradation are just a few of the consequences of "the dominant patterns of production and consumption" that have emerged through unequal development (thereby calling for new rules that resist the old economic rules that harm people and planet). To untangle that suffering, humans must commit to changing institutional and social ways to be in global relationships and to create life-giving solutions. The preamble ends with a nuanced explanation of how each person on earth has responsibility for the present and future, in local and global communities, and must live in solidarity with all life, with gratitude and reverence. This preamble sets up the sixteen principles as the foundation of a global ethic of community.

The first section, "Respect and Care for the Community of Life," includes the first four principles. It begins with a call to respect all life in its diversity and dignity (including the language of "intellectual, artistic, ethical, and spiritual potential" of humanity).[5] It differentiates levels of responsibility, saying that with more power comes more responsibility for the common good. That power and good mean that human rights and freedoms should be established in communities that are, among other things, peaceful and just—and must be sustainable for future generations. In this section we see a commitment to the diversity and rights of all parts of life on earth, both present and future.

The second section, "Ecological Integrity," includes four principles around themes of protection and preservation. Protection of endangered species and systems, control of non-native and genetically modified organisms, management of renewable resources, minimizing the overuse of nonrenewable resources—each of these actions allow for the "special concern . . . for the natural processes that sustain life." In this section, too, the precautionary principle is lifted up as a guide: that a concern for the long-term effects should govern decision-making in all aspects of life. In doing so, restraint in use of materials, energy, and technologies becomes an avenue to fair and just economics, universal health care, and knowledge and resources in the public domain. Economies of care are better than economies of dominance.

The third section, "Social and Economic Justice," draws upon the commitments of the first two sections with four principles that connect social justice to creation justice. These principles call for the end of poverty, education for all people, empowerment of all humans with education, and the re-centering of the needs of people who live in marginalized communities. Corporations and international trade are called on to be transparent and bridge the gap between the rich and poor. Specifically, the empowerment of women and girls is connected to sustainable development; women need to be allowed to participate as equal partners in all aspects of decision-making. This section ends with a commitment to the end of all discrimination and to the access of all to "human dignity, bodily health, and spiritual well-being."

The final section, "Democracy, Nonviolence, and Peace," explores further the connections between governance and ecological justice in four principles. Here the right to informed participation in government and the right to freedom of expression are upheld for citizens, and the responsibility of institutions and governments to be uncorrupted and active in environmental care is articulated.

Access to education of all kinds (formal and informal, scientific and artistic) is articulated as a lifelong right, as well as the need to see spiritual education as integral to sustainable living. All animals—both domestic and wild—are to be treated with respect and dignity. Finally, the principles call for a culture that privileges peace over conflict, exploration over exploitation, and wholeness and solidarity over disputes and splintering of communities.

The charter ends with a reiteration of human interdependence with each other and with all parts of life on earth. This interdependence means that global dialogue and decision-making must continue with "a change of heart and mind." Living out the principles of the Earth Charter requires a commitment to making hard choices that center marginalized communities and a reverence to life. These choices get easier as we remember that we belong to each other. And these choices require a change in the rules, especially public policies and the rules that govern corporate life, that contradict the principles.

While I have summarized many of the principles in binaries, many of the principles refuse easy binaries, slipping connections to education and gender in among commitments to transparency in governance, and repeatedly suggesting that demilitarization is helpful to the promotion of several principles. The principles, while calling for an integrated response to the particular time we humans find ourselves in, are well-integrated and are themselves interdependent.

Whenever I read the Earth Charter, I'm reminded of the age of the document. The process of writing the Earth Charter began in 1992 and took ten years to complete, written by many different people. Its birth in the early 1990s reminds me that we have known about the groaning of creation—and who is at fault for climate destruction—for so many years, and that humans have been struggling to change the course of environmental degradation for decades. On the one hand, I find so much hope in that struggle—how can we *not* be encouraged by a document written by thousands of people? It has had years to affect policy and life together, its principles seeping into how humans rethink their relationships to each other and to all parts of the earth. On the other hand, I am discouraged that humans have been trying to reimagine and articulate the need for different ethics in response to a time when "the world becomes increasingly interdependent and fragile [and] the future at once holds great peril and great promise" for more than thirty years. How have we *not* been changed by this document and this peril?

(That is a great question! I would say that environmental racism plays a huge part in that. —Michael

Right, because the powerful won't change for marginalized people, even when the marginalized are many. —abby)

The Earth Charter reminds me that humans have already been trying to articulate a new narrative with new rules that could and would make us live together on this earth responsibly and with attention to our interconnectedness. The charter's articulation of a new narrative (or perhaps it is a new articulation of an ancient truth) and its principles or rules remind me that whatever way forward we go, we go together. Maybe that collective narrative and its life-supporting rules will not be enough to stop climate change, but it will be part of changing who we are to face the new realities of a climate-changed world. As we tell that new story, with rules that feel new but are rooted in the old story of the early church, we can begin to live into that story and what could be. Now, in practical terms, how are we to do that?

Changing the rules of public policy is one pathway for living into that new story and moving toward what could be. Public policy—be it local, national, or international—is perhaps the most influential form of rules determining how our life together is set up. Public policy determines who will have clean water, whether fossil fuel companies keep up their relentless extraction, where toxic waste will be dumped, whether solar energy will be funded for marginalized communities, whether taxation and land zoning will support or thwart small Black-owned and Indigenous-owned farms, and so much more. The Earth Charter principles are worthy guides for public policy and that is a part of their purpose.

We share now a few stories of people who are striving to build new rules for economic life, rules that align with the Earth Charter principles. These examples cohere also with the early church's commitment to set up the rules of life together so that resources are shared and none go without. Our examples are local, national, and international.

Local

First a local example, from Michael: Environmental racism has caused us to "other" others. In addition, society has us set up to think that our selfishness is independence. We tend to do what makes us comfortable. We are very rarely concerned with the effect that it will have on someone else as long as it does not negatively affect us. I first came across North Birmingham through my friend and brother Michael Hansen. Michael was the executive director of Greater Birmingham Alliance to Stop Pollution (GASP) in Birmingham, Alabama. Michael is a fierce and passionate leader for justice. While I was at his office when I first moved to Alabama, he explained to me the conditions of North Birmingham, a community of towns in Alabama.

North Birmingham is the home of a superfund site and some of the most egregious disrespect from manufacturing and the government that you would ever want to see.

When we drove up, I noticed that the roads looked slick, as if it were sprinkling, but it was not sprinkling and had not rained. Michael explained to me that this was the residue from the coal firing steel plant. There is a warning for people to not allow their children to play outdoors without a time limit. They have told the people not to plant anything in the yard because of contamination.

These communities have been in bad shape for a long time. Generation after generation has been trapped watching their loved ones die prematurely. The mayor of Birmingham is from one of these communities. You would think that public officials would have mitigated this issue; they have not. When I started bringing attention to this and the work of GASP to help find relief, I was amazed to learn that some communities in Birmingham had no idea this was happening to their neighbors. Part of the ignorance was due to the media being controlled by the industry that is destroying these communities. The other part of the ignorance is that we are not concerned about anything that does not directly affect us.

Through the organizing of GASP in North Birmingham, GASP was able to expose the corruption that Boch and Bingham, Alabama Power, and ABC Coke used to hinder the 35th Avenue superfund site from expanding. GASP provided a platform to amplify the voices of community members that countered the narrative expressed to the public on their behalf by the polluters that were poisoning them. Changing the narrative was direct and done at a grassroots level to bring attention to the issues. This attention has led to partial relief for some of these community members. I use "partial relief" because this is an ongoing fight that community-led organizations like PANIC (People Against Neighborhood Industrial Contamination) are still leading. This is an example of how changing the narrative and amplifying voices helped to change the rules on a local level.[6]

Local organizing, together with mobilizing on a national level, is even moving the international needle. Consider, for example, the People's Justice Council's Fair Shares Collaborative. This collaborative has birthed the Shift US Campaign. The campaign aims to inspire the US—our citizens, institutions, and government—to do its fair share of global climate action. We mobilize at the local level to advocate at the federal level for climate action on the global level. This is not only a grassroots campaign, it is a moral call. We write:

> The United States is the largest historical emitter of greenhouse gas emissions globally and has the greatest capacity to reduce emissions as a country. Therefore, it stands to reason that the US should do the most to combat the climate crisis: Not only bringing down its own emissions immediately but also helping other developing nations leapfrog technologies and behaviors that emit emissions that are not necessary if we help finance and share technology and resources. That is the US's Fair Share of solving the climate crisis. Is it happening? No, but the Fair Shares Collaborative, which is facilitated by People's Justice Council and has dozens of US and International partners, is showing a pathway that our country and communities can take to create a better world for all of us.[7]

From a moral perspective, the US has a hefty price to pay for its share in the damages it has and continues to do to the planet. Many people refer to repaying this climate debt as climate reparations. The US must not only repair the damage that it has caused (to the extent that that is possible), it must compensate for the losses that have been incurred.

In developing the Shift US Campaign, we understood that doom and gloom messages were not working. Guilt-ridden messaging also has ill effect. What does work is a hope-filled message of limitless bounds. By providing spaces at the local level to imagine and build collectively, the Shift US Campaign is providing the tapestry for our collective future. In essence, the world we want is the world we create. The only requirement is that it is unlike the one that we currently live in. It should be a world that sees its obligation to life as sacred. It should be a world that sees, hears, and gives weight to all voices. It is a world that centers care over commerce. It puts people and planet over profit. It appreciates the experience of one another individually and makes decisions communally. It is the world we want; it is the world we can build. However, it requires a shift in collective conscious. Shift US provides a platform to dream of and advocate for the world we want. A shift in collective conscious will change the rules because the people will recognize their collective power and demand change.

Through policy, the Fair Shares Collaborative put together and introduced the Fair Share NDC. The Nationally Determined Contribution is basically the share that each country self-determines to pay toward tackling the climate crisis. This was established at the Paris Climate Agreement in 2016. The Fair Share NDC was written and introduced to the United Nations at the Conference of the Parties 2024. This effort was led by Natalie Lucas,[8] an incredible leader and organizer. It was built out and championed at the local grassroots level. The process of putting it together was inclusive, and its impact has been powerful. The influence of the Fair Shares NDC can be seen in the language of the recent NDC submission by the United States. This is an example of how we are changing the rules through messaging, organizing, and mobilizing at the local and national grassroots levels. It also shows how we center care throughout our movement spaces.

National Level

Nationally, Third Act has been organizing Americans over the age of sixty to "change the world for the better."[9] Started in 2021 by Bill McKibben and many others, Third Act is led by a volunteer advisory council and organizes through working groups that are determined by geography or theme.[10] All working groups are committed to the same guiding principles:

Be kind
Be inclusive
Boost others
Be humble
Take care
Back up the youths!
Be generous, but not to a fault!
Be accountable
Be creative
We're in this together[11]

These principles, at odds with much of United States culture, guide working groups as they seek to change the rules. Working groups organize around Third Act's major campaigns safeguarding our climate and democracy, and around aligned local priorities. Major campaigns aim at changing the rules of what is acceptable in our time. Campaigns include short-term actions like calling on Congress to unfreeze federal funds or calling on Costco to stop partnering with Citibank, which funds fossil fuel devastation in Louisiana and in other parts of the US. Other Third Act campaigns are long term, working for fossil-free finance and supporting voter rights. These are efforts to change the rules so that people and corporations with power and governments are more accountable to communities. You can join them, wherever you are in the US, by checking out the list of actions at https://thirdact.org/act/.

International Level

The Fossil Fuel Non-Proliferation Treaty Initiative was begun by grassroots organizers in the Pacific Islands who saw that the international community and governments were not being held accountable to the goals of the Paris Accords and other climate justice measures. The initiative calls for "a concrete, binding plan to end the expansion of new coal, oil, and gas projects and manage a global transition away from fossil fuels."[12] Sixteen nation-states have worked together to demand new rules and laws that would hold fossil fuel companies and industrialized governments accountable to countries that are already suffering the effects of climate change.

The treaty calls for three things:

1. Stop building new fossil fuel infrastructure
2. Develop a fair phaseout of existing fossil fuels, with most developed countries going first
3. Support a just transition to new energies, without leaving workers behind.[13]

The communities organizing around these three parts of the Fossil Free Non-Proliferation Treaty note that "our planet is at a crossroads. . . . It's time for governments to take commensurate action, stop throwing fuel on the fire and join the bloc of nation-states seeking a Fossil Fuel Treaty negotiating mandate to help build a safer world for all, free of fossil fuels and powered by abundant, renewable energy."[14]

These are the kinds of new rules that we need so that people around the world can flourish. These new rules would be legally binding, meaning that (unlike the Paris Accords) countries would have to follow through on their commitments. Communities at many different levels around the world are organizing around the treaty, including churches and denominations, synagogues and mosques, cities and states, universities, and other organizations. You can too. Learn more about the treaty and take action at https://fossilfueltreaty.org/.

Invitation for Action

In order for us to distance ourselves from those who do not see or care, we will need to intentionally center care for our underserved communities. What does not affect us directly affects us indirectly. We can center care through real and practical work to change the world by changing the rules.

How can you do that? So much concrete work can be done to change the rules of public policy by linking with an organization that is working to do that locally, nationally, or internationally. We've already introduced you to Third Act and the Fossil Fuel Non-Proliferation Treaty Initiative. You could also explore (and join) the work of your state's League of Conservation Voters and learn how they're building power around racially inclusive and environmentally sustainable political policies: https://www.lcv.org/.

State affiliates of Interfaith Power and Light (IPL) are organizing as people of faith around climate and environmental issues. Many state affiliates host Lobby Days, when people of faith meet with state officials and policymakers to ask them to write and support legislation that is rooted in racial and environmental justice.[15] Find your IPL affiliate at https://interfaithpowerandlight.org/affiliates/. Likewise, many Christian denominations have public policy advocacy offices. You might contact yours and join your faith community in its efforts to change the rules of public policy as a dimension of caring for the neighbors whom we are called to love and the earth that we are called to serve and protect.

To learn more and to meet a global community of companions, be sure to visit the website for this book series at https://www.buildingamoraleconomy.org/. Select "Book 1: Climate." There you will find a treasure trove of resources—organizations and

websites, reading material, film, and more—to accompany you on the faithful journey toward climate justice.

Blessing

One of my favorite poems is by the theologian and pastor Martin Niemöller, "First They Came For. . ."

> First they came for the socialists, and I did not speak out—because I was not a socialist.
> Then they came for the trade unionists, and I did not speak out—because I was not a trade unionist.
> Then they came for the Jews, and I did not speak out—because I was not a Jew.
> Then they came for me—and there was no one left to speak for me.
>
> —Martin Niemöller[16]

My prayer and blessing is that God helps you to see, hear, and feel others. I pray that God permeates your heart with so much love that you would need to give it away to maintain balance. I pray that when you speak up for the least and the left out, God gives you power to speak truth and break chains of oppression. May you be blessed in your fight.

CHAPTER FOUR

Move the Money

WE HAVE BEEN exploring how climate change, ecology, faith, and morality are interconnected. We looked at the realities of climate change and environmental racism, and we showed that new rules are possible. One way we build the new and change the rules is by moving money out of the systems that create climate change. This is the fourth finger on the hands of actions we can take to build a moral economy: Move the money.

In this chapter, we review how climate justice organizers have used divestment and reinvestment as a tactic to create a new economy for climate justice. Since 2012, organizers have used this tactic (learning from other divestment movements) to create a new economic possibility for people of faith. I (abby) will write about my faith community, which spent ten years organizing for divestment from fossil fuels and voted to move our money out of Exxon, Chevron, Marathon, Mobil, and Shell in 2022. I argue that this move has meant that our denomination takes its faith commitments seriously in relationship to our denominational money and investments. In this chapter, I review the tactic of divestment, how it's been used in two other social movements, how several denominations have used it in relationship to fossil fuels, and how the Presbyterian Church (USA) used it to divest from these five companies.

This is a big chapter, long because the divestment from fossil fuels movement represents a lot of what these first three fingers of action call us to. Michael will end this chapter with a reflection on how the economic model that is shown in the book of Ruth has resonance with the divestment from fossil fuels movement. He reminds us that after divesting, we must invest in solutions that center our beloved community. Then we end with an invitation for action and a blessing for the work.

First, a word about the work of money: while some argue that the value of money is the ability to make more money, many sociologists would add that money often plays a key role in meaning- and identity-making. These two understandings of money, of course, are not mutually exclusive. Christian biblical traditions affirm this second understanding in the Gospel of Matthew, which says, "For where your treasure is, there your heart will be also" (Matthew 6:21, New International Version). As such, investments and spending show what issues and concerns are important to investors. That is, economic choices are moral and meaning-making choices.

Divestment is one way a moral economy emerges. Divestment deploys an economic mentality that uses the selling of investments and products as a way to hone corporate profits and mission and emphasizes the meaning-making role of money in social movements and religious communities.[1] To explore that effort, we begin this chapter by viewing how divestment as a social movement tactic has been used by people on the frontlines of dismantling apartheid in South Africa and genocide in Palestine by Israeli leaders.

The Role of Divestment

Divestment has been used in social movements as a nonviolent tactic to apply public pressure to organizations, corporations, and nations to align investments with beliefs, drive conversation, and draw attention to social injustices. Individuals and institutions practice divestment by removing their investments in a company from their stock portfolios, doing so alongside public statements about why they no longer will profit from stock of that particular company or industry. It mimics the economic ideology that uses divestment to "help identify and highlight the boundaries of an organization."[2] Corporations have long used divestment as a means of focusing on corporate mission, and thus economic decisions are moral decisions. Similarly, social movements have used divestment to name and shape the values and commitments to which a particular group of people adheres (or does not). It is one way for people who do not live in the context of oppression to stand with those who do. Since the financial market and institutions are global entities, it is also a way to insist that the international community pay attention and insist on liberation. It is one way to live into an understanding that the humanity of one part of the world is bound up in the humanity of all others.[3]

Divestment does at least two things in a movement: First, it calls on institutions and individuals to align their values and their money in a framework in which money is not only a capital or currency but also a reflection of an individual or institution's values, a way to show where support for values or beliefs lies. Second, it supports the momentum of the larger movement of which divestment is a tactic, essentially becoming a spotlight to shine light on an issue. Thus, engaging in divestment shows the opinion of an individual or institution about a particular social justice issue.

Divestment was a tactic in social movements used on an international scale in the eighties and nineties to force the dominant white South African government to end its system of apartheid.[4] The decades-long struggle led by exiled, banned, imprisoned, and frontline Black Africans and allies called upon individuals, universities, companies, churches, and other organizations to show their support by taking their investments out of companies based in South Africa. This call was used to escalate pressure on the apartheid government, adding to demands from athletic boycotts, sanctions by the United Nations, and in-country protests to show that the international community supported the anti-apartheid movement. The fact that Archbishop Desmond Tutu was one of the major voices and faces of the call for divestment from South Africa added to the moral impact of that call and every subsequent use of divestment in movements.[5] This includes the Boycott, Divestment, and Sanctions (BDS) movement, which targets Israel, and the global divestment from fossil fuels movement. Hundreds of institutions have responded to these calls for solidarity.

Even before the anti-apartheid movement began, investments were used to establish meaning; money is never neutral. *(Amen and a finger snap on this! — Michael)* In fact, in his work on the "ethics of investment," Charles Powers commented that divestment, "the concept, if not the term, is responsible for the fact that the portfolios of many institutions (especially [some] churches) have never included stock . . . from tobacco or liquor."[6] That is to say, corporate institutions with endowments have long operated with considerations about the ethics of where their money is and how they make it matter. In the same way, faith-based institutions have long interpreted their Christian public witness as impacting their financial stewardship, spending, and investments. The PC(USA) is just one example of a Christian denomination that has made investment choices and then later divested from companies doing business in apartheid South Africa and Israel.[7] This framework of money as meaning-making and public witness emerges in social movements through economic sanctions, which include boycotting and shareholder engagement as well as divestment.

Other Economic Tactics

Boycotts and Shareholder Engagement

Boycotts usually involve refraining from purchasing physical products from or doing other kinds of business with a particular person, company, or institution. The term "boycott" was coined in 1880 when Irish tenant farmers and townspeople banded together to refuse to work with or associate with Captain Charles Boycott after he delivered eviction notices to the tenant farmers who could not pay rent in a depressed economy.[8] Similar collective economic action happened in the Revolutionary War in the US, for example, when colonists refused to buy British tea.[9] Other boycotts have included "the boycott of Nazi Germany that was promoted by some Jewish organizations following the German elections in March 1933," which was "an early example of a boycott that was aimed at a state on account of its policy of racial discrimination."[10] Each of these examples demonstrates how boycotts have to be selective because boycotts "have been promoted by social movements and subaltern groups with limited resources and capacity."[11] Indeed, that the people who organize boycotts have limited power in comparison to the economic entity they seek to change is one reason they must organize and mobilize through a shared, targeted tactic.

The ethos of shareholder engagement is based on the concept of staying at the table to shape the company, whereas the ethos of divestment is based on the idea that staying at the table suggests that those at the table approve of the work of the company; if one disapproves, then one should leave the table. Shareholder engagement uses the ability of anyone who holds stock in a company—and thus has a voice and vote at shareholder meetings where decisions about the company are made—to vote on, and potentially pass, resolutions about the identity and the business model/plan of that company.

Shareholder engagement was used for the first time in the anti-apartheid movement in 1971 by Paul Neuhauser and the Episcopal Society for Cultural Racial Unity (ESCRU) in partnership with other faith communities that created the Interfaith Center for Corporate Responsibility.[12] This was because the South African government was ignoring the sanctions, embargoes, and protests against them (more below), and companies operating within South Africa were participating in apartheid through discriminatory employment practices. Neuhauser believed that as a stockholder in General Motors, the largest employer of Black South Africans, ESCRU could make a demand that General Motors change their practices.[13] In response, Reverend Leon Sullivan, a civil rights leader in the United States, was elected to be the first African American on the board of General Motors. He then developed the Sullivan Principles, which were first adopted at General Motors and then by other corporations that had operations in South Africa. These guidelines called for better employment practices like equal pay and the end of segregation in whatever corporation signed onto the principles. It is important to note that the principles included, as Behar points out in *The Shareholder Action Guide*, a clause about moving from shareholder engagement to divestment: "If these actions did not bring about the end of apartheid in ten years, Sullivan called for all companies [that had agreed to the principles] doing business in South Africa to divest their operations in the country, thereby putting a time limit on the process of reform for these companies."[14] Now we will look at how divestment has been used in response to climate change.

Divestment and Climate Change

Divestment as part of a larger movement is a way to push a particular identity and value system into the world. While there are financial implications for the companies and stockholders, often the primary point of divestment in a movement is to make a moral statement, not necessarily to affect the bottom line of a company, industry, or country. Divestment is an affirmation that the institution creating a problem cannot continue business as usual, slow reformation is not enough, boycotts can only go so far, and deliberation or negotiation can dilute ethical commitments.[15] But how does divestment as a moral witness happen? Powers outlines how owning and selling investments match the beliefs and values of an organization. He divides them into negative sanctions—"a decision to sell stocks in protest over practices or products" and "a decision not to buy stocks in such companies until particular practices are discontinued or initiated"—as well as incentive investments—"a decision to look with special favor upon the stock of [some] corporations."[16]

Institutions and individuals that seek to divest can do so in two broad ways: categorical and company-by-company divestment. Categorical, which has been used in the fossil fuel divestment campaign, means that there are no funds invested in any company in the fossil fuel industry as a sector. A company-by-company strategy, used in both the calls for divestment from South Africa and Israel, means that investors look at each company on its own and decide if they want to sell or keep holdings in the industry.

In 2012, *Rolling Stone* published a piece by Bill McKibben about three important numbers regarding climate change. In "The Terrifying New Math," he painted a bleak future caused by human-made climate change rooted in science and what he called "our precarious–our almost but not quite finally hopeless—position."[17] After the article was released, McKibben went on a national tour, teaching and speaking about a movement to divest from fossil fuels, just as activists had called for a movement for divesting from South Africa during Apartheid. While McKibben initially targeted secular organizations like universities and municipalities, faith-based organizations and denominations quickly began to join the movement to divest from the fossil fuel industry. These denominations and churches joined the divestment movement to add their moral and religious voices, asserting that profiting from the industry that has been the largest contributor to the creation of climate change was against their theological and biblical beliefs.[18]

McKibben's "The Terrifying New Math" was one of the first instances in which the fossil fuel industry was called out in a popular publication. The article was accessible and clear to a general audience, and it made connections to the successes of the anti-apartheid movement to use a collective moral voice to respond to an immoral action. It was a call that McKibben has repeated in the following years.[19] And it's a call we must continue.

In short, the article said that we must stay below adding two degrees Celsius to our global temperature, that to do that we can put only 565 more gigatons of carbon dioxide into the atmosphere, and that at the time there were 2,795 gigatons of carbon in the reserves of fossil fuel companies. McKibben was quick to say that previous strategies to work with the fossil fuel industry failed to adequately mitigate climate change and thus environmental activists needed to see the industry for what it is: the enemy. Environmentalists, anyone who cared about the planet, and anyone who was dependent upon the planet had to turn the sense of the enemy into anger—and then into a movement.

In the US tour that followed the article, McKibben called for divestment from fossil fuels, using the language that "if it is wrong to wreck the planet, then it's wrong to profit from that wreckage." This language of wreckage was taken from "The Terrifying Math" and incorporated into the messaging of all three sub-movements, and it clearly connects moral witness and financial investments.

McKibben writes that the point of divestment is to make a moral statement that eventually leads to vital change rather than to dramatically affect the bottom line of the fossil fuel industry:

> The idea is not that we can bankrupt these companies; they're the richest enterprises in history. But we can give them a black eye, and begin to undermine their political power. That's what happened a quarter century ago when, around the western world, institutions divested their holdings in companies doing business in apartheid South Africa. Nelson Mandela credited that as a key part of his country's liberation, and Desmond Tutu last year called on all of us to repeat the exercise with the fossil fuel companies.[20]

Here we encounter the historical component of the movement to divest from fossil fuels, particularly how divestment was used to fight apartheid. In my conversations with the founders of Fossil Free UCC, Fossil Free PC(USA), and Fossil Free UMC, each pointed to McKibben's article and the subsequent Do the Math Tour as a catalyst for the movement in their denominations. Jim Antal, president of the Massachusetts Conference of the United Church of Christ and founder/director of Fossil Free UCC, said that to him, McKibben "sounded like a trumpet blast from Ezekiel." Antal and McKibben had been arrested twice at the White House protesting the Keystone XL pipeline, and the three days they spent in jail further clarified for Antal that "God is calling the church to a new vocation" in response to the reality of climate change.[21] Likewise, Dan Terpstra, one of the founding members of Fossil Free PC(USA), said that the work of Fossil Free PC(USA) started with McKibben's Do the Math Tour:

> About the time the Do the Math stuff was starting up, the University of Tennessee Knoxville campus assigned *EAARTH* [McKibben's 2010 book on climate change] to all freshman and he was going to speak and I immediately asked him to come to [my church in] Oak Ridge. I told him I was coming to the Forward on Climate March for Feb 2013. . . . It was at that rally that I attended a breakout meeting at an Episcopal church and one of the conversations was on divestment and I said that I wanted to work on exploring getting the Presbyterian church to divest.[22]

For Jenny Phillips, the founder of Fossil Free UMC, the article was a catalyst for more conversations. She was struck by the article clearly naming the enemy, saying that "the enemy is the fossil fuel industry and we should divest." She went on to say that she wanted the church to divest, and she started talking with the people around her to figure out where their church money was going. They were stunned to realize so much of it was in fossil fuels.[23]

Each of these conversations was initially isolated in its own denominational context. Soon after the article's publication, GreenFaith, an international and interfaith environmental organization, helped create faith-based language for the movement. Its executive director, Fletcher Harper, convened several conference calls, wrote a white paper for use by sub-movements and 350.org (the secular organization that McKibben founded), and created comic books. Key were those conference calls, which allowed a variety of voices to come into the conversation with their own gifts, expertise, and perspectives. In our interview, Harper said that there were "people like Bob Massie, an Episcopal priest who was active in the apartheid movement and had really close relationships with the ICCR, as well as Tom Van Dyck, a financial advisor at the Royal Bank of Canada, who was the first fund manager to make a strong and principled case for divestment. They added a level of financial credibility [that] you can and should do this and you should fire your manager if they say you can't do this."[24]

By collecting these voices, GreenFaith and Harper hoped to create an ecosystem of voices that could make the case from a variety of backgrounds. Harper and GreenFaith provided the groundwork for a movement that reached across denominational lines. GreenFaith sought out people who could answer questions that came up around divestment, and they, in Harper's words, "established the credibility of the argument."

These conversations were happening in the larger context of the secular movement to divest from fossil fuels, which had begun to gain momentum in college campuses and municipalities from 2012 onward. The denominations were crafting specific faith-based language.[25] By 2013, when the UCC voted to divest, over three hundred universities and one hundred cities had divestment campaigns.[26] At this point, it is helpful to tease out the stories of each of the denominations that have been working on divestment from fossil fuels, specifically looking at the individual stories of their movements.

The United Church of Christ and the Divestment from Fossil Fuels

In our interview together, Antal talked about how he wanted the movement in the United Church of Christ to be a success story and a model for other denominations. Indeed, it is often faith communities that can "take the risk of new behavior, and its success becomes the story to be told to others."[27] And so, the

UCC was the first denomination in the United States to vote in favor of divestment from the fossil fuel industry. Their work is deeply tied to McKibben's.[28]

While Antal wrote the first resolution on divestment, between the Massachusetts vote and the General Synod vote, ten other conferences voted either to cosponsor or endorse the resolution.[29] The work of organizers was rooted in theological, biblical, and scientific rationale. Antal preached on divestment and organized several congregations to support the resolution. Since that vote, the United Church of Christ has kept tabs on the rest of the movement and developed options for fossil-free investments for congregations and individuals. Antal noted in his interview that divestment work has now become part of the work of the Office of Environmental Ministries for the denomination and that he has often has an intern specifically for divestment work.

Presbyterian Church (USA) and Divestment from Fossil Fuels

For the Presbyterian Church (USA), the beginning of Fossil Free PC(USA) was a grassroots effort. Rebecca Barnes, who in 2012 was the national staff person for Environmental Ministries in the PC(USA), describes the beginning of the sub-movement in the PC(USA): "I was impressed that this was the first time [I could remember] that individual Presbyterians on their own set up phone calls, took notes, took leadership, and didn't depend on a staff person. . . . It was exciting because there were Presbyterians who were galvanized about an issue of climate change that gave them energy."[30]

In Barnes's description, we see parallels with Nepstad's description of movement emergence. There must be a favorable political climate for the activists to perceive that they will be successful, which we see in Barnes's description of the energy and noncontroversial identity in the beginning of the movement. There has to be a preexisting organization to support the movement, which we see in Barnes's institutional support of the movement by convening phone calls and connecting people. Nepstad writes that it is necessary, too, for people to "no longer consider the status quo legitimate. They begin demanding change, and they believe that they have the power to alter the situation."[31] This last condition emerged in the PC(USA) after decades of shareholder engagement with corporations. In the years since the movement for fossil fuel divestment has worked through the PC(USA), members of the committee that handles shareholder engagement have particularly lifted up white supremacy and racist capitalism, and the need to center people of color.[32]

In 2022, the Presbyterian Church (USA) voted to divest from Chevron, Exxon Mobil, Marathon Petroleum, Phillips 66, and Valero Energy, the five companies that scored the worst in the criteria created by the church's Mission Responsibility Through Investment Committee. By the end of 2022, the

PC(USA) had divested its holdings from those companies as a denomination. After ten years of organizing, Fossil Free PC(USA) successfully won a divestment vote, albeit on a smaller scale than delegates had asked for.[33]

The United Methodist Church and Divestment from Fossil Fuels

Jenny Phillips rooted the beginning of Fossil Free UMC in the Do the Math Tour and the subsequent secular divestment campaigns. Those campaigns were asking their governing boards to divest from the Carbon 200 list.[34] Phillips and others quickly realized that that would not happen in the UMC because the denomination's investment decisions were made by values, not companies. And so, Phillips and other Fossil-Free UMC activists drafted a resolution to General Conference, the denomination-wide gathering, that mimicked the demands of university campaigns but included faith-based language. In 2014, four annual conferences (the yearly meetings for regional bodies in the UMC) studied the resolution and in 2015, twelve annual conferences out of fifty-six in the United States voted on whether to bring the resolution to General Conferences. Eleven of the conferences affirmed the resolution and one ended up not voting. In 2015, Phillips worked with activists in twelve conferences to prepare them for General Conference.

In return, Fossil Free UMC faced significant resistance. Phillips said that when they began their outreach to delegates to General Conference, Wespath, the pension board for the UMC, began to push back, saying that as a pension board they "didn't want to be constrained." Phillips attributed the nimbleness of Fossil Free UMC in their ability to quickly respond to Wespath's negative responses, which included a publicly available video opposing divestment. Wespath's responses are another example of how institutions feel threatened by activists create obstacles.[35] In 2016, General Conference voted against divestment, but since then Wespath has set up options for fossil-free investments.

In a time of global climate crisis, economic disparity, and social disconnect, communities of faith must find ways to grapple with and dismantle systems of oppression and suffering to be relevant in the public sphere. One way faith communities demonstrate their religious leadership is through economic solidarity. The fossil fuel divestment movement in the UCC, PC(USA), and UMC shine as case studies in faithful responses to climate change. Each of these sub-movements has raised the level of dialogue in its denomination about the urgency of climate change and the fiduciary responsibility of denominations to respond morally. Divestment is not just about economics.

In 2016, the Center for International Environmental Law wrote in in their "Trillion Dollar Transformation: Fiduciary Duty, Divestment, and Fossil Fuels in an Era of Climate Risk" that "it is increasingly clear that climate change and climate risk are already reshaping the investment landscape, and that these effects will grow dramatically in the years ahead."[36] The report goes on to say that money managers need to include climate change in their assessment of risks. The conversation has changed; removing funds from the fossil fuel industry is no longer just a consideration for environmental and religious activists.[37] As dialogues about climate change and divestment have happened in a variety of sub-movements, it has been important to continue to frame those conversations in terms of moral and religious language, not just economic vocabulary. From the beginning, McKibben has followed Archbishop Tutu's lead to say that divestment is not an economic issue but a moral one.

There are, of course, structural reasons (like a history of colonization and white supremacy) why white people and Global North institutions have access to wealth and assets and have then been targeted to use that structurally unfair access to wealth to be in solidarity with communities that have been marginalized or are calling for solidarity through divestment. In the US, divestment has been a vehicle for majority-white denominations because of their historic wealth, linked to the white supremacy, colonialism, and privilege behind such wealth noted above. They have thus been called upon by movements and grassroots leaders to use that economic wealth in solidarity.[38] These grassroots movements often have been communities of color who are the most impacted by injustice, demanding that white communities do something in response to that injustice.[39] Divestment from social injustice, and subsequent reinvestment in just funds, is thus a strategy that uses capital to articulate values and reallocate funds. That's part of what it means to build a moral economy that recognizes climate change in the intersecting realities of capitalism and white supremacies. But it's not enough to recognize these issues. We need to move the money, rooted in the stories of our faith that shape us.

Invitation for Action

"Put your money where your mouth is!" When I was growing up in East Atlanta and hanging with my peers and family, I would hear this saying frequently. It meant, if you believe in it, invest in it. This proclamation would normally be followed with some type of action. Either the listener would retort, invest, or get quiet.

We'll dive into the story of Ruth in the Bible more in the next chapter. But: When Boaz sees Ruth working in the field, he is left with a choice of putting his money where his mouth is. He would go on to invest because he believed in his role as kinsman redeemer. (It helps that Ruth was fine as frog hairs!) It was because he believed in what he was investing in that he gave it his time and resources.

The opposite should happen when we *don't* believe in something. When we do not believe, we should not support it by investing in it. Boaz had a cousin who was the rightful kinsman redeemer. He could have been the one to marry Ruth. Boaz gave his cousin an opportunity to assume the role. However, the cousin declined. The explanation that we are given is that the cousin is concerned with the peace in his home. Perhaps, however, because he did not believe, he did not invest. I hope you are seeing how this works.

Capitalism throws rocks and hides its hands. It does harm and never takes responsibility. A report called "Dirty Pearls: Exposing Shell's Hidden Legacy of Climate Change Accountability, 1970–1990," reveals that the company Shell Global structurally developed in-house knowledge about global warming.[40] They knew that their activity was contributing to climate change, but they did not warn the public. Warning about the oncoming crisis would have decreased profits. So, the information was buried and we are where we are today.

Our call to action is simple, seen, and warranted. This climate crisis—including the devastation and disruptions caused by storms and wildfires—is not by happenstance. The fossil fuel industry is doing terrible damage to God's creation. Therefore, as stewards of God's creation, we must act.

An individual divestment would not really register in the grand scheme of things. When we suggested that you get a better bank at the end of chapter 2, we noted how this individual action could spur you into *collective* divestment action. If we divested not only as individuals but also collectively—as faith communities, institutions, cities, and other groups—from those that harm us and God's good creation, we could make the fossil fuel industry become a major contributor to the solutions needed rather than being a major contributor to the crisis we are in. The answer is simple; we must divest from the harm and invest in care for God's creation and created. We must center care. How can you do that?

Here is a first step:

As you ask yourself how you can center care in your investments (if you have them), begin to consider how you're investing in the world. Do you have a 401k or 403b? What funds are being used? Start to ask your financial planner or investment officer questions about who they support, from a place of curiosity. As you start to ask these questions, and move your money as you are able, you start to live into the triple bottom line, an economy that values people, planet, and profit. Ask them to move your money to funds that support the beloved community.

And here is a second:

Ask yourself: What groups—be they large or small—am I a part of that have investments? My church? The national denomination of my church? The college, university, or seminary that I attended? My city or neighborhood association? Is there already a divestment campaign afoot in it? How could I find out? (You might check out the Global Fossil Fuel Divestments Database at https://divestmentdatabase.org/ to find out about divestment campaigns in groups similar or related to yours.) What likeminded people might I invite to join me in convincing this group to center care and build the beloved community through its investments? What joy might we find in living out our faith in this way?

To learn more and to meet a global community of companions, be sure to visit the website for this book series at https://www.buildingamoraleconomy.org/. Select "Book 1: Climate." There you will find a treasure trove of resources—organizations and websites, reading material, film, and more—to accompany you on the faithful journey toward climate justice.

Blessing

The work for climate justice is long and laborious. There is no easy fix, no simple solution.

In the long work, remember that you belong to your community.
May you know your community. Know it in the ways that you know your family—with all the good and the bad. May you know what is needed and where there is more than enough. May you find collaborators for the struggle, coconspirators who long for climate justice and are unafraid by the future to come.
May you be brave as you move the money, investing in justice and connection.
And may your courage draw you closer to God, in faithful love.
Amen.

CHAPTER FIVE

Grow the Bigger We

TAKE A DEEP breath. In the preceding chapters, we've argued for how the unequal reality of climate change—rooted in white supremacy and exploitative capitalism—requires those of us who are Christian to change. (We are well aware that this call also pertains to people in other faith traditions and no faith tradition, but at the moment we two Christians are calling on Christians.) We have to build new economic and financial systems that value people and planet. We have to chart out new rules and policies that will govern these ecological systems that shape the use of money and other resources. We have to move our money, using how we spend and invest it, individually and institutionally, to help shape a more just world. Each of these actions is connected to growing the bigger we, the fourth finger on the hands of action. In growing this bigger sense of who we are together—the "we"—we go about the work of healing exploitative economies in favor of life-giving and collaborative economies.

We do this rooted in faith and in the stories that can still sustain us. In this chapter, Michael shares more about his story and explores how the book of Ruth can help root us in a larger story that reminds us we are not just ourselves but are part of a larger community that deserves life-giving economies. The book of Ruth is an old story in our faith, and returning to it reminds us that the work of healing is often about returning to what roots us. (And if we are really calling you to be more radical people of faith, this is perhaps where we drop one of abby's favorite facts about word origins: The word *radical* is from the Latin for the word *root*, or what is grounding/beginning/centering. So, as you feel pulled to expand or radicalize, keep breathing. You're really just coming home.)

Speaking of home, here's Michael on how he grew up: Receiving a Pentecostal Holiness upbringing in Atlanta, Georgia, was not for the faint of heart. When you think of being a closed society, you likely glean a great deal from my upbringing in the church. At almost fifty years old, I am still learning how to be open and inclusive. In my church, if you did not believe and act as the dominant culture, you were often condemned and ostracized. My experience in this setting was that it was not about pleasing God as much as it was about not displeasing those who were influential (not always the clergy or leadership).

While earning my Master of Divinity from the Interdenominational Theological Center (ITC), I met Rev. Susan Mitchell, who in turn introduced me to Rev. Tim Downs and the United Church of Christ (UCC). The UCC was the opposite of what I grew up with. Through the ITC and UCC, I found a community centered around inclusion and care. Juxtaposing the two experiences, I find that being exclusive creates burden. However, being inclusive is beneficial. *(Oh my gosh, say that! —abby)* A deeper look at the book of Ruth further exposes us to the benefits of growing the bigger we.

Learning from the Book of Ruth

A term comes up in Ruth that I believe is that book's theme. This term is *hesed.* Hesed is one of those Hebrew words that is not describable in a single word in the English language. Hesed is expressed in relationship. It is the loyalty, faithfulness, kindness, goodness, mercy, love, and compassion that may be experienced in a relationship. Although it can be seen in human relationships, it is hard to find. However, it is clearly seen in the relationship between God and God's creation. What is important in this chapter is that you know that these attributes may be practiced in our lives and that you recognize hesed as a guiding light for building networks, alliances, and other forms of collaboration to heal exploitative economies in favor of life-giving and inclusive economies (growing the bigger we).

(Experienced in real life, in real community. It's funny that language can only get us so far. At some point, we have to be in the real, practical world. —abby)

In this chapter we first explore growing the bigger we by examining the book of Ruth. Then we consider more closely how what we learn from Ruth can inform growing the bigger we today. In the first section, I use the encounters and relationships of the characters of Ruth to further engage the hesed in the text. Just as my experiences in relationships shaped my theological, sociological, and economical perception, I believe that the call for hesed in the text should shape our theological, sociological, and economical view. Through Ruth and my experiences we will learn that the way we see God and engage in community is shaped by our experiences. To experience more of God and to grow the bigger we, we must be open to experience relationships with others. (Recall that when we say "grow the bigger we," we are talking about building networks, alliances, and other forms of collaboration to heal exploitative economies in favor of life-giving and inclusive economies. And we are talking about building relationships of inclusion, reciprocity, care, and equity.)

Introduction to Ruth

We cannot ignore Ruth's ties to the book of Judges in the Bible. Ruth opens with, "In the days when the judges ruled. . ." The author wants you to know that this story takes place during the judges era of the story of the people of God. In order to understand the importance of Ruth, you have to understand the mess of an era that we call "judges." The era of the judges can be summed up in two words: chaotic and inhospitable. It was a time when God's chosen were a chosen mess.

One theme of the book of Judges is that the people sinned against God. Another theme is their subjugation at the hands of others, which meant that they were always at war. Then, they cry out to God. The last theme is that God is faithful to faithless people. *(God is always faithful and the people—us included—are always a little chaotic and messy. —abby)* This is the cycle of this dysfunctional relationship. The book of Judges is a canon of relationships gone wrong. These horrible relationships left them in a cycle of chaos, destruction, and deliverance. The book of Ruth breaks through the chaos in the era of the judges as a respite. It provides for us an example of relationships gone right.

The author of Ruth is unknown. Some scholars argue that the purpose of Ruth is vague. I cannot agree with this. I believe that the purpose of Ruth is to teach us the benefits of the bigger we. When juxtaposed to the perverse relationships in the book of Judges, this purpose becomes very clear. While the need to conquer and dominate causes conflict and constant repentance in the life of Israel, forming inclusive bonds based on honor and loyalty as seen in the book of Ruth provided benefits in the life of Naomi, Ruth, Boaz, and the broader community.

(I love too how the book of Ruth centers on the life of women and how they are faithful to who they are meant to be. In the preaching commentaries for Ruth, there's often mention that God is a secondary character. But isn't God part of hesed? And so isn't God at the heart of the book?

Ruth goes with Naomi. I always wondered what happened to Orpah, who lost everything and then lost more when she took Naomi up on the invitation (the demand?) to leave. To whom did she return? What kind of community was left for her? What became of her "we"? In no way do I judge Orpah. I just wonder what happened to her. —abby)

The Benefits of the Bigger We

Ruth's refusal to leave Naomi's side in Ruth 1 is the first time we are introduced to hesed. It is seen in Ruth's loyalty to Naomi. Ruth is from Moab. When Elimelek and family left Bethlehem, Judah they settled in Ruth's land. When life started life-ing for Naomi, she was in Moab.[1] When Naomi left Moab, she left with Ruth.

Ruth's willingness to journey with Naomi in life meant that an elderly widow would be comforted. Ruth was still young and fertile. Ruth's husband had died, so she could remarry and start life over. Another aspect of hesed is that it is performed for a situationally weaker person by a situationally stronger person. When the bottom had fallen out and life was life-ing for Naomi, Ruth was left with the same choice as Orpah. Ruth could have left Naomi in her vulnerable state. Orpah turned back to her place of comfort. She chose her own comfort and self-interest. *(To be fair, maybe Orpah was trying to be an obedient daughter-in-law to Naomi. —abby)* Ruth chose to journey with, care for, and comfort Naomi—her elder. One benefit of growing the bigger we is that vulnerable people are valued, included, and cared for.

Another benefit of growing the bigger we is that power and resources are likely to be more equitably distributed. Capitalism's emphasis on profit views everything through the lens of commodity and consumerism. Since America's birth, America's history has been of slavery, genocide, and oppression. We govern and decisions are made based on profit, not people or planet. People are valued through the lens of property, labor, and entertainment. Resources and power are concentrated in a few hands rather than shared. This should be seen as an offense to God.

Growing the bigger we does not take advantage of positions of power or use them to consolidate wealth and more power. Growing the bigger we shows mercy. This is another way that we see hesed being done. The actions of Boaz in relation to Ruth illustrate this. Boaz was a person of great influence. He had land, workers, and title. Ruth was a sojourner who had been reduced to gleaning the leftovers in the field. Gleaning is an important concept to lift up because it shows an economy that included care for the vulnerable. We see hesed in the honor bestowed upon Ruth during Ruth and Boaz's first interaction. Boaz bestowed this honor upon her when he said that he had heard about how loyally she had been in caring for Naomi, his kin. Boaz expressed appreciation to Ruth for her doing hesed. His appreciation of Ruth led to her acceptance into the community.

Ruth and Orpah are attributed as doing hesed. Ruth and Orpah were Moabites. This means that they are outsiders. If you are keeping tabs, hospitality and loyalty: Moabites 2, Israelites 0. It is the "other" in the text who introduces us to attributes of God in God's relationship with us. It is through Ruth that we see care and concern coupled with commitment. Ruth, this foreign woman who was not supposed to be married to an Israelite because she was an outsider (see Ezra 9 and 10), becomes the matriarchal ancestor to King David. Hesed was practiced, the we grew bigger, and community and well-being expanded. That expansion is another benefit of growing the bigger we and of practicing hesed. Now, let's take a look at hesed in practice in the present day.

Learning from Contemporary Visionaries

I would like you to meet an excellent leader and visionary named Collette Pichon Battle. Collette is the cofounder and codirector of Taproot Earth. I remember talking with Collette on the phone. She asked me to join a call for a coalition that she was putting together called Gulf South for a Green New Deal. She told me that she was calling on the entire Gulf South to mobilize for climate justice. Her vision was aspirational and solid. I was hooked.

We started meeting and expanding. Each call offered new faces and motivational words by Collette. Before you knew it, we were raising money and realizing the vision. What stands out to me about the process is how broad she made the call to action. I was not the only one to receive a personal invite. Before long, we were inviting others. What she started began to grow organically because we had buy-in. We had taken part ownership of Collette's vision. We were champions of climate justice in the Gulf South. *(So many different voices and experience coming together to work for community, over 350 organizations. Learn more at https://www.gulfsouth4gnd.org/. —abby)*

The Gulf South for a Green New Deal was developed by practicing hesed. It was developed relationship by relationship. The Gulf South for a Green New Deal eventually became Taproot Earth. The formation grew into a global organization with members from around the world working to achieve climate justice from a grassroots level. The benefits from the relationships under this umbrella grew beyond the boundaries of the Gulf South. Here again we see the benefit of expansion when we grow the bigger we through practicing hesed.

We expand methods, modes, and impacts when are introduced to varying perspectives in growing the bigger we. The Black Panther Party was established in 1966 by Bobby Seale and Huey Newton to combat police brutality in Oakland, California. Realizing the interconnection between poverty and oppression, they established community services to care for their community. One of the programs that they established was the free breakfast program. This program ensured that children in their community did not go to school on an empty stomach. Schools offered reduced lunch, but not free lunch for poor children. The Black Panther established the first free lunch program in this nation because it wanted to ensure that those on the outer margins were included in care. This program was adapted by the federal government, which now offers free lunches to those on the outer margins. A program started by this organization of Black justice activists to grow and care for the bigger we was established and expanded to an even larger we.

Invitation for Action

There is so much work to do to build community—the bigger we—across differences. But we are not without examples, as we see above from both the Biblical and contemporary witness. And we have schools of thought and practice, like ecowomanism, to guide us even further. Melanie Harris describes ecowomanism as "an interdisciplinary approach to environmental ethics that centers the lives, experiences and perspectives of women of color and women of African descent."[2] That is, the discipline seeks to decolonize environmental ethics from the white gaze (meaning controlled by people who are white) and embrace the voices of others, not simply white or Black perspectives, while working for justice for women and for earth. It is committed, then, to the expansion of environmental discourse outside of the white voice as the norm, which is what we've been trying to do in this book. Harris writes that "acknowledging the profound source of wisdom and morality that religion, spirituality and social protest faith communities have been and are for people of color, [an ecowomanist] perspective considers the cosmologies that ground these faith systems as primary sources from which to reinforce earth ethics."[3] Her discussion of the foundations of ecowomanism is grounded in the environmental justice movement, African American religious thought, womanist thought, and ecofeminism.

How do I (abby) as a white ecofeminist—and how do you—learn from such a foundation? And how can each of us be moved to action by it? How and what do we learn about growing the bigger we through practicing hesed? First I must be willing to acknowledge that the "horrors suffered by the people of color in the face of climate change, and other environmental catastrophes, the devastating loss of their homes, and the humiliating ways their sacred and spiritual connection to the earth has been treated" are real.[4] It is not fiction that people of color and people who are poor (social categories that often—but not always—overlap) suffer from the effects of climate change first and worst.

Ecowomanism reminds us that we humans are indeed resilient. It reminds us of the generational trauma that parts of humanity have already suffered. If we believe that Black and Brown bodies have been the sites of trauma—and here I mean environmental as well as racist, such as the toxic waste industry's practice of strategically planning dump sites near minority and poor communities—what does their survival tell us about the humanity capacity to survive trauma?[5] How can you engage with these questions today? Make a list: Whose voices am I listening to that reveal realities of climate injustice and other environmental hazards faced by people of color and economically impoverished people and realities of the economic conditions of their lives? Do my sources include voices from these communities? In light of ecowomanist values and insights, to what new sources will I shift in order to include such perspectives? And whose voices could reveal for me alliances and networks aimed at countering that injustice and building more equitable and climate-friendly economic practices and polities?

To learn more and to meet a global community of companions, be sure to visit the website for this book series at https://www.buildingamoraleconomy.org/. Select "Book 1: Climate." There you will find a treasure trove of resources—organizations and websites, reading material, film, and more—to accompany you on the faithful journey toward climate justice.

Blessing

Breathe in.
The prospect of our annihilation traumatizes us.
And climate change does that to us on a grander scale.
We must find hope on a grander scale.

Breathe out.
In Christianity, we talk about the resurrection of Jesus as God doing the most impossible thing.
In the resurrection of Jesus, we understand that God is more powerful than human forces and that God makes a way—lighting a way when the way seems unclear.
It reminds us that God is present in creation around us.

Breathe in.
Breathe out.
Resurrection is for us all.
Just as Ruth said to Naomi—I will belong to you,
Just as Boaz said to Ruth—I will belong to you,
We have to say to all creation—I will belong to you.
And in doing so, we build the bigger we.

CHAPTER SIX

Resist the Wrong

MAYBE YOU'RE GETTING to this chapter and wondering if it will be possible to make the changes we need to make: transform or create new financial systems, change the policies that shape our systems, move money and rely on each other. These are hard because they require us to resist what is all around us, to resist the wrong rules that pit us against each other, extracting from our labor and the planet more than is reasonable or right. Faith in Jesus, who turned tables in the Temple when the money changers were taking advantage of the rules and who taught a mutual aid that resisted imperial economies, reminds us that it's possible, faithful, and necessary to resist. Our faith makes us strong in this work. In this chapter, we share a story from the resistance to a pipeline and dig into how our faith can be a throughline of support as we resist what is wrong.

Faithful Resistance

In mid-2024, climate activists in Tanzania who were campaigning with GreenFaith were harassed by their government. Their phones were taken and destroyed, and local police insisted that the organizers travel from their home villages to police stations multiple times to face questioning. This travel often involved great distances and costs. The approximately 120-kilometer trip by motorcycle cost more than forty US dollars round trip and six hours on the road each time.

The ten community members, a group including people of faith, had been organizing to stop the East African Crude Oil Pipeline (EACOP), a fossil fuel pipeline funded by France-based TotalEnergies. The pipeline had sunk the villagers deeper into poverty by taking their lands, and it had disrespected the burial places of their loved ones and infringed on their basic human rights.[1]

The mostly Christian organizers in Tanzania had been working alongside organizers in France who identify as Muslim, as well as traditional leaders, local administration, and local media, to disrupt the building and use of the pipeline. It would cut across Uganda, Tanzania, and Kenya, without properly compensating local communities for destroying their ecosystems or collaborating with communities in ways that honor their way of life and their spiritual, social, economic, cultural, and environmental needs. Indeed, the Tanzanian activists charge that the destruction of these ecosystems and the ongoing extraction of fossil fuels violates their religious beliefs and their rights, destroys their communities, disrespects family graves, and displaces peoples, all without respectful community engagement, consultation, or following legal or traditional due process to solicit input and consent from the communities most impacted. According to East African activists in the "As If Nothing Is Sacred," a report coordinated by GreenFaith:

> Most East African families keep family burial sites near their homes, so that they can fulfill traditional and religious requirements. When families must relocate their household due to EACOP, they may lose access to their family graves. Project officials have provided little to no support for the proper relocation of graves closer to the family's new home. This makes it impossible for these families to fulfill their responsibilities to their deceased family members.[2]

In May 2024, GreenFaith organized an international fast for people of faith to be in solidarity with the Tanzanian activists who were harassed.[3] In June, GreenFaith released an open letter signed by people of faith from around the world calling on Samia Suluhu Hassan, the president of Tanzania, and Patrick Pouyanné, CEO of TotalEnergies, to halt both the harassment of the organizers and the construction and use of EACOP.[4]

This series of actions in multiple countries by people of many faiths illustrates one way in which people of faith respond to climate crises by resisting what is wrong—and it is a beautiful example of growing the bigger we as discussed in the previous chapter. The situation also illustrates the ill-treatment and push-back committed by corporate interests, in conjunction with governments, when people act on their concerns regarding climate change and their beliefs to care for the planet, its people, and places. This is but one example of religious climate activism to resist the wrong, and it calls to mind other well-known resistance movements such as at Standing Rock in the US or the Alberta Tar Sands in Alberta, Canada. Real stories of resistance by people of faith who will not let the world suffer without a struggle are out there.

Loving Each Other as Resistance

At the core of Christian faith is the commitment to love one another. And that love can sustain us as we resist amid the worsening global climate crisis. Although there are notable people of faith who claim that climate change is not real (like Calvin Beisner and others at the Cornwall Alliance), or that it isn't predominantly human-caused, there is no meaningful debate about the science of climate change.[5] There really is not. Climate change creates a devastating and deadly reality, especially for people around the world who have been marginalized.[6]

Throughout 2023 and 2024, global heat waves broke records.[7] These heat waves have caused deadly drought and loss of food crops, contributed to famine, and impacted religious traditions. As David Wallace-Wells reported with accounts from around the globe, these heatwaves led to an unprecedented number of heat-related deaths. He noted that during the 2024 hajj, the pilgrimage of Muslims to Mecca (one of the five pillars of Islam), the official count was over 1,300 deaths.[8] Essentially, rising heat impacts the ability of Muslims to fully practice their faith.

Most religious traditions have ethics of care for other humans, and many people of faith connect their religious ethics with the deadly realities of climate change. Then they seek to respond in ways that demand change. They are heeding the religious call to love one another, especially those who are suffering.

These faithful responses have emerged in many forms, including as resistance to climate injustice and the policies and practices that create it. abby's research on how people of faith have organized around divestment from fossil fuels reveals a relatively new dimension of religious ecological activism—the growing numbers of rituals and resources for lament and despair. How can we hang on to grief and resiliency in the face of an ongoing struggle for climate and environmental justice?

Ways of resistance are all around us. Faith-based communities and nongovernmental organizations bring generations of responding to crises and trauma in general, and they have begun to create those resources for this particular crisis. In a February 2023 blog post, for example, Reformed Christian writer Debra Rienstra wrote, "We [people of faith] should know all about facing our fears with faith and partnering with God's redemptive work toward a sweeping vision of flourishing for all people and all the earth. We should know about healing and repair and community."[9]

And Dayenu, a US-based Jewish organization, describes part of their work as supporting "people to confront the reality of the climate crisis and to grapple with the deep spiritual and existential questions it raises. Rooted in Jewish wisdom, tradition, ritual, and voice, we provide pathways for people to confront and move through feelings of fear, anxiety, and powerlessness."[10]

People of faith can utilize bold practices of faith and action while also modeling grief. This spectrum of faith-based responses, including more urgent and challenging ones, is needed to face climate changes and challenges while also countering the despair that studies show is gripping so many.[11] In an ongoing time of climate change, perseverance will be needed. There are difficult decisions ahead, and not everyone will recover.

The Role of Lament in Resistance

The reality of inequality in survival that the world has seen more clearly since Hurricane Katrina is rooted in racism and capitalism—sinister sisters (or maybe cousins, as Michael says) that have harmed people who are poor and people of color. It is one dimension of the wrong to be resisted. For example, we can work to build awareness, public policies, and corporate regulations that enable people of color and poor people to recover from and be protected from climate disaster to the same degree more climate-privileged people are.

In *The Body of God: An Ecological Theology*, Sallie McFague theologically situates the state of the world in lament. She creates a systematic theology of God's body on earth, articulating how humans have refused to live in healthy relationships with one another and creation; where and how God interacts with creation; and how Christ and salvation impact creation and particularly the parts of creation that are oppressed.[12] Faith groups can and do bring resources like people power, places to meet, money, and access to social connections. They can craft rituals and resources that name the space between what needs to be done, what is being done, and what simply will not or can no longer be done. They can offer us tools for lament. In an article in *Sojourners* on climate grief, Chelsea MacMillan, an activist and interfaith minister with GreenFaith, says that her approach to leading people in a climate courage group is to "practice *via negativa* and *via positiva*, as complementary experiences of God." She begins with the *via negativa*, letting participants name loss with questions like, "What's breaking your heart?" or "What are you holding?" Then, she centers gratitude during *via positiva*, asking "What support do you have in your life?" or "What do you have to offer?" She closes with a song."[13]

Maybe lament exists as ritual or space-keeping. But it might also look like taking action, as the Tanzanian Ten take proud action in response to EACOP. José Gonzalez-Colon, who supported divestment as the moderator of the PC(USA) Synod of Puerto Rico, was one of the thirty Presbyterians who walked 260 miles in two weeks across the Midwest to raise awareness of the need for the denomination to divest from the fossil fuel industry. In an interview, he expressed sentiments that were common among the walkers. He wondered how best to present the lament for being unable to fulfill our Genesis-based vocations. Several times in my interview with him, Gonzalez-Colon wondered how we could frame the divestment movement through our human vocation to care for the planet through the lens of caring for people who are marginalized, saying that "money has to follow the ethos of our faith." He later said, "We have a religion of consumption and capital and social stratification. . . . And we have to start facing this hand in hand with how we address climate breakdown. We need to look at the people who are oppressed and where they are on this matter and how they have been bringing prayers and rituals of lament."[14] Lament and resistance, for Gonzalez-Colon, go together. Both call us to connect to frontline climate communities and hear from them.

We, each of us on this planet, need people of faith more than ever in the climate movement as we move into lament because faith communities already have language and rituals of lament, and community connections to enable communal lament. We need lament as part of the movement because we are in such dire straits. If we really face the reality head-on, it's easy to despair. Still, this lament can fuel our resistance. We can boldly say: We grieve this good and beloved planet. We grieve how the powers and principalities of the world have extracted fossil fuels and more from the earth. We will not be silent, even as we grieve.

More Stories of Resistance and Lament

There are so many stories of people of faith and goodwill resisting, together, the wrong of climate injustice. In this chapter we have encountered a few. Take a moment now to savor a glimpse of a few others.

Rise St. James is a faith-based environmental justice organization working in Cancer Alley in Louisiana. They've been organizing to keep petrochemicals and plastic plants out of their community. They root their work in education (including teaching about a "chemical of the month") so that activists show up ready to argue their perspective. Under Sharon Lavigne's leadership, they're resisting the wrong assumption that polluting industries are welcome in their minoritized communities.

Earth Quaker Action Team (EQAT) has been resisting fossil fuel finance by regularly protesting the investment firm Vanguard, which is based near EQAT's headquarters in Pennsylvania. Sometimes they sit in rocking chairs, sometimes they sing and chant and pray. They keep showing up, in faith, resisting the wrong investments of Vanguard and demanding new investments in community.

Little Village Environmental Justice Organization, based in Chicago, was founded by public school parents to resist dangerous chemicals in a school renovation. They've expanded their struggle to successfully close power plants in, and add more green space to, their mostly Latino neighborhood. As a community-based organization, they've been resisting the wrong policies that have allowed polluting industries to take up shop in their communities.

All of these communities are working to resist the wrong. They're demanding that we collectively invest in beloved community and all that is good. These are powerful stories. They must be told and heard.

These stories are especially compelling in contrast to the story of how we could have stopped climate change before it got so bad. This failure is what breaks my heart. In 2018 Nathaniel Rich published an article in *New York Times* called "Losing Earth: The Decade We Almost Stopped Climate Change." A year later his similarly titled book was published.[15] Both pieces recount how scientists and activists asked the US Congress to respond to the specter of climate change in the 1980s, before everything got beyond our control to solve it. They brought out charts and facts and figures—the best science at the time. James Hansen, a leading climate scientist, testified on one of the hottest days on record at the time, June 23, 1988. Congress failed to enact any meaningful legislation. The first time

I read Nathaniel Rich's piece, I cried. I lost sleep. I despaired. If people with a lot of power couldn't change the tide of climate change, what could I do? The second time I read that *New York Times* piece, I threw myself into the grass on the farm where I was living. On my hands and knees, I tried to remember to breathe. I lay down and looked up at the sky: I let the earth hold me as I wondered—what to do? It was my own moment of lament and grief for the planet and our human failure to hold fossil fuel companies accountable. That lament and grief is what I come back to in my own resisting: Oh, how much pain, suffering, and death could have been avoided if people with power had done the right thing when they could.

Invitation for Action

Stories of resistance abound. What stories of resistance do you know about? Who were the leaders? What sustained them? How did they make change? What are the places of resistance where you need to show up? If you are a parent, grandparent, aunt, uncle, or another guide for young people, how will you model for them that resisting the wrong is a dimension of faith and of faithful love for others?

Many people of faith are learning faith-based community organizing (also known as broad-based community organizing) as a means of building beloved community. The arts of community organizing equip people for resistance in its diverse forms. They also are marvelous tools for building the bigger we. What about learning from the Community Organizing for Climate Justice as Love in Action program of the Center for Climate Justice and Faith? According to its website, the program "seeks to equip leaders with a climate justice framework and practical tools to implement team-based climate resiliency projects in their communities. A three-day online training followed by 12 weeks of weekly coaching and leadership development will allow program participants to understand themselves as leaders, expand their capacity to enroll others in collective action, and understand the roles of teams in creating change."

Learn more and consider signing up with a team from your community at https://www.plts.edu/programs/continuing-education/community_organizing_climate_justice.html.

To learn more and to meet a global community of companions, be sure to visit the website for this book series at https://www.buildingamoraleconomy.org/. Select "Book 1: Climate." There

you will find a treasure trove of resources—organizations and websites, reading material, film, and more—to accompany you on the faithful journey toward climate justice.

Blessing

This is a blessing of lament, for when you need permission to let your heart break over all that could have been.

Take a deep breath.

Remember that in the beginning, God looked at creation with love. It was very good.
Remember that God made all of creation out of topsoil.[16]
And when God made humans out of that same topsoil, God breathed life into us.

So, breathe with God.

As you breathe, think of what you love about creation. A plant, a tree, a person, an animal, a community.

What do you love about this part?

How is that part of creation affected by climate injustice? How is that part of creation exploited? And if not yet, when will it be?

Let your heart break over this exploitation.

Take a deep breath.

Hold an image of that part of creation that you love in your mind. A plant, a tree, a person, an animal, a community.

Let your heart break and also let yourself be angry.

It did not have to be this way. (Real people decided to not make the world better.)

We live in a world that could have been more connected. (Real people broke us apart.)
It once was connected.

And yet: that part of creation that you love, a plant, a tree, a person, an animal, a community, will be transformed and harmed by climate change.

Take a deep breath.
Let your heart be broken.

May your broken heart lead you to action, resisting all that is wrong.
Amen.

CHAPTER SEVEN

Change the Story

In this chapter, we're digging deeper into how society's overarching stories impact us personally. To make a more moral economy, we must recognize and critique the stories that shape us, our economies, and our society in the United States. Some of these stories have damaged the earth and all that relies on its ecological systems; some of these stories have been deadly. Changing the story, the next finger on the hand of action for a climate just economy, means we have to reclaim more life-giving narratives. In this chapter, we start with Michael's story and how he has experienced the interconnected sins of capitalism and racism. Then we move into how abby has interpreted the white sin of climate racism (by which she means that climate injustice is linked to white supremacy, and that is a problem for white folks to deal with) in conversation with some biblical stories that shape our faith. We end with a call to action and a blessing. Be gentle with yourself in this chapter. Listen to how you feel your body respond to these stories and to how white supremacy and advanced capitalism have damaged so much. Your gentleness is a way to resist that wrong story and reclaim a better one.

Michael's Story

Capitalism is evil and racism is economical! Growing up in Atlanta is a different southern experience. Atlanta is considered the Black mecca of the South. Having traveled throughout the southern parts of the United States of America, I see clearly how distinct the experience of life in Atlanta is. I grew up surrounded by Black excellence. However, I was not isolated from experiencing the harmful effects of capitalism and its demonic twin, racism.

White supremacy makes no sense. It is the most irrational shit that I have ever come across. Capitalism is the only thing that benefits from racism. Yet we continue to participate in both. White supremacy is made up to protect white mediocrity and the economic status of white property-owning males while killing people and the planet. Yet we continue to participate. *(Often when we know better. And just to underline what you've said: White supremacy doesn't even protect most white folks. —abby)* Make it make sense. As a Black man, it is hard for me to understand the rationale and mechanics behind racism and white supremacy. I am not the creator or driver of these diabolical monsters. I am often a victim of its deployment. Therefore, I can only tell you how they have informed my being in society.

James Baldwin wrote that "to be Black in America is to be in a constant state of rage."[1] Racism is something that I experience rather than enact; reliving difficult encounters also brings up old feelings, like rage. I am also struck by the reality that nothing has changed. Capitalism is destructive. We can see it in real time. The wealth gap is widening, our communities are polluted, wars and genocide are being funded while justice and inclusion are being fought. Our seniors and single mothers are stuck in a cycle of poverty while trying to keep their lights on and their homes climate-controlled. These are effects of capitalism. *(We make Black folks poor and then force them into horrible jobs with long hours; they come home to houses that are not easily heated or cooled. I want to know why we are so easily convinced that it is the fault of poor people and folks of color that they are poor. —abby)* Racism and white supremacy are still being perpetuated. When I reflect on capitalism, racism, and white supremacy, not only am I enraged but I am also frightened in knowing that the things I have endured, the generations before me will also endure.

In this chapter I write from my experience and my rage. I will not hide how capitalism, racism, and white supremacy have made me feel. I will attempt to show you what white supremacy and racism feel like through my lens. No worries; I will include other voices. I can also assure you that my coconspirator dr. abby will jump in on the discussion. *(Already sitting with you, Rev. Taking a breath in gratitude for the story you tell.)*

My Beginnings

I encountered racism at birth. I was born in 1975 in Atlanta, Georgia. I was born to working-class parents. During my time in the home, we were a family of four. My father had two sons who would occasionally live with us, too, making us a family of six sometimes. While I was in the military, my mother and father had their Abraham and Sarah experience: my baby sister was born in 1995. My upbringing was filled with joys and struggle. I had the joy of being in a large family. Still, we struggled because we were a large family that fell in between services that could have assisted my family's upward mobility.

Atlanta is not really Georgia. It is the South and not really the South. Atlanta is Atlanta! In our home we struggled with utility insecurity. We did not have stability, so we moved around a lot. We often lived in rental homes or lease-to-own homes. What I would later come to understand is that it is expensive to be poor. And, racism is economic and systemic, which means that there is a whole lot of support for it to keep going. When you are poor and Black, it is hell.

(It's a hell I can't imagine as a white person, because racism backs white people with legal and social power. —abby)

As Robin DiAngelo writes in her book *White Fragility*, "When a racial group's collective prejudice is backed by the power of legal authority and institutional control, it is transformed into racism, a far-reaching system that functions independently from the intentions or self-images of individual actors."[2] *(This means that while your family was full of good intentions to provide the best life possible for each other, there was only so much your parents could do inside systems to change the economic reality of your family. —abby)*

In the book *Critical Race Theory: An Introduction*, Richard Delgado and Jean Stefancic write that there are two camps of critical race thinkers. One camp is the idealist. They write that the "idealist" camp:

> holds that racism and discrimination are matters of thinking, mental categorization, attitude, and discourse. Race is a social construction, not a biological reality. . . . Hence we may unmake it and deprive it of much of its sting by changing the system of images, words, attitudes, unconscious feelings, scripts, and social teachings by which we convey to one another that certain people are less intelligent, reliable, hard-working, virtuous, and American than others.

The contrasting camp is the realist. The "realist" camp, they say:

> holds that though attitudes and words are important, racism is much more than a collection of unfavorable impressions of members of other groups. For realists, racism is a means by which society allocates privilege and status. Racial hierarchies determine who gets tangible benefits, including the best jobs, the best schools, and invitations to parties in people's homes.[3]

I would say that I am a hybrid of both of these camps. I hold that race is a social construction, not a biological reality. This is why I say that white supremacy does not make sense. I also hold that racism is much more than a collection of unfavorable impressions of members of other groups. Racism is systemic and economic. The history of Atlanta housing shows that racism is systemic and economic.

Racism Is About Economics

In 1922, Robert Whitten designed the zoning ordinance for Atlanta that segregated whites to the R1 section and Blacks to the R2 section. *(I just learned more about zoning laws in Atlanta—R1 is for single-family homes and R2 is for residences with more than one family and for commercial. So essentially, Whitten set the stage for white folks getting to have a home of their own. —abby)* The Atlanta Planning Commission explained that "race zoning is essential in the interest of the public peace, order and security and will promote the welfare and prosperity of both the white and colored race."[4] *(Meaning that race-based zoning is essential for white folks to do better than others. —abby)*

This preemption of redlining followed the development of the Harding administration's Advisory Committee on Zoning. Hoover was a staunch segregationist, and his commission had staunch segregationists on it. This committee shaped federal policy that segregated communities. In 1924, the national real estate board adopted a code of ethics that included these words: "A realtor should never be instrumental in introducing into a neighborhood . . . members of any race or nationality . . . whose presence will clearly be detrimental to property values in that neighborhood."[5]

Redlining was an all-hands-on-deck operation to economically oppress Black upward mobility. By implementing redlining laws and racist community covenants, Blacks were located in areas that were zoned for commercial use. Rothstein states, "I think it can fairly be said that there would be many fewer segregated suburbs than there are today were it not for an unconstitutional desire, shared by local officials and by the national leaders who urged them on, to keep African Americans from being white families' neighbors."[6]

(It feels really important to underline that this impacted Black folks economically. It also impacts white folks. My gut reaction to this quote is a reminder of my own white childhood that included Black neighbors. Mr. Jenkins and his granddaughters were mainstays in our lives—we played and swam together. Mr. Jenkins tended his garden just across the fence from my parents' garden, and I remember passing vegetables back and forth. My life was and is better because of their friendships. I don't raise this as a "look at me, I've always had Black friends" note. I raise this as "look at this! Everyone *is better when we are neighbors to each other." —abby)*

This racist plot continuously profited off the labor of Black people. This was not only a move to continue building wealth, it was also part of a longer move to protect stolen wealth. Cedric J Robinson writes:

> In the year 1877, the signals were given for the rest of the century: the Black would be put back; the strikes of white workers would not be tolerated; the industrial and political elites of North and South would take hold of the country and organize the greatest march of economic growth in human history. They would do it with the aid of, and at the expense of, Black labor, white labor, Chinese labor, European immigrant labor, female labor, rewarding them differently by race, sex, national origin, and social class, in such a way as to create separate levels of oppression—a skillful terracing to stabilize the pyramid of wealth.[7]

We felt it where I lived. In my neighborhood in Atlanta, we have single-family homes that share space with gas stations, liquor stores, and plazas. We are fewer than three miles from the highway. The roads and drainage systems have been neglected. Anytime there is heavy rain, streets are flooded because trash and debris clog the drains. The county has said that it is not their responsibility to clean the drains and has put the responsibility on affected homeowners. Although the homeowners are not being serviced, they must still pay taxes.

This is what racist capitalism looks like. The previous illustration is indicative of my life growing up Black in Atlanta. In my youth, places like Buckhead, Avondale Estates, Marietta, and Alpheretta were off limits to families like mine. We would rarely venture to those places. We would definitely not find ourselves in those parts after dark. Let's just say that we did not feel welcomed.

But another reason that we were unable to access the suburbs is because of the lack of transit. In 1971, the transit compromise happened. This compromise limited the transit system to the inner city. Marta, the Atlanta transit network, could not extend to the suburbs. Marta was limited to Dekalb County, Fulton County, and the city of Atlanta. This was also where the largest concentration of Black people were, which meant that Black people who needed access to public transit to move could not get out to the suburbs. The transit system was limited by racism.

In 1970, Black people made up 21 percent of the metro Atlanta population. They were concentrated in a swath of land east and west of downtown. White residents comprised 76 percent of the population. They lived throughout the ten-county metro region, except in areas with high concentrations of Blacks.[8] Restricted transportation meant restricted access to jobs and upward mobility. Redlining still limits our access today. Those apartheid communities that existed then are the communities that suffer from environmental racism today.

Environmental racism is real. Dr. Robert Bullard states, "The South has always been thought of as a backward land, based on its social, economic, political, and environmental policies. By default, the region became a 'sacrifice zone,' a sump for the rest of the nation's toxic waste."[9] An example of this would be the Arrowhead Landfill in Uniontown, Alabama. This landfill sits in the middle of this historic Black community. It has accepted coal ash from the Tennessee Valley Authority (TVA). This coal ash was deemed too toxic for the predominantly white and affluent Kingston, Tennessee. However, it was it was livable for predominately Black and rural Uniontown, Alabama. When I spoke with one resident of Uniontown, I learned that she and her husband moved back to their family's land in Uniontown to build their dream home. She said that the landfill has made their dream a nightmare.

What I need you to see is that the government backed and implemented racist institutions through public policy. The lack of investment in infrastructure, limited access to upward mobility, and degradation have robbed us of wealth. *(Governments doubled down by putting pollution into those places. —abby)*

Race is a social construct that emerges from the pit of hell. This makes white supremacy weird. By "weird" I mean: White supremacy is the belief that white people constitute a superior race and should therefore dominate society, typically to the exclusion or detriment of other racial and ethnic groups.[10] *(This belief is all the more dangerous because in many cases it is assumed subconsciously. Many white people do not realize that—on some unrecognized level—they hold this belief and are influenced by it. That makes it more dangerous because you cannot confess or fix something if you don't acknowledge that it exists. And even if we know that we may hold these racist ideas and want to unlearn these ideas, it's so ingrained in us that we can repeat or enact white supremacist ideals without meaning to. —abby)* I mean, if a stranger walked into a room with no proven evidence claiming that they were superior to everyone in the room, wouldn't that be weird? This is what white supremacy looks like to me. White supremacy is dangerous. White supremacy wants to dominate society and does not care who it harms to do so.

You ever heard of the United Daughters of the Confederacy? They were founded on September 10, 1894. They were founded in Nashville, Tennessee, and were introduced on a public platform in, you guessed it, Atlanta, Georgia. Their main purpose was to preserve and romanticize the Confederacy through the promotion of the Lost Cause, which argues the cause of the Confederate states during the American Civil War were just, heroic, and not centered on slavery.

Here were the goals of the United Daughters of the Confederacy:

1. To honor the memory of those who served and those who fell in the service of the Confederate states.
2. To protect, preserve and mark the places made historic by Confederate valor.
3. To collect and preserve the material for a truthful history of the War Between the States.
4. To record the part taken by Southern women in patient endurance of hardship and patriotic devotion during the struggle and in untiring efforts after the War during the reconstruction of the South.
5. To fulfill the sacred duty of benevolence toward the survivors and toward those dependent upon them.
6. To assist descendants of worthy Confederates in securing proper education.
7. To cherish the ties of friendship among the members of the Organization.[11]

The United Daughters of the Confederacy spun the narrative of Southern rebellion by romanticizing stories of the South. I remember their propaganda being in our textbooks in the early eighties. I grew up in East Atlanta. In my community was a confederate statue. Guess who's responsible for this constant reminder of trauma? You guessed it: the United Daughters of the Confederacy. The Daughters of the Confederacy shaped how society looked at the South and the role of Black people in it. Their propaganda and media like *The Birth of a Nation* promoted racial stereotypes and white supremacy. It was all made up; it was sin!

I am more of a realist than an idealist. I do believe that we have to face the truth and see that racism is systemic. The idealist in me also advocates for changing the narrative. If I blend the two ideas, I come to the conclusion that to be an effective change agent, I must speak the truth and change the narrative.

abby and the Sin of Climate Racism

As we move into the second section of this chapter, we focus on sin and white supremacy in the narrative of climate change. We shift into abby's voice because it really needs to be the labor of white folks to dismantle the white sin of white supremacy. We humans are made in the image of God and called to live in relationship with God and each other with "mutual openness and help."[12] Sin emerges when we break those relationships, when we break the web that connects all of us. Sin happens when we forget to love others and forget that we ourselves are loved. When we need to change the story, we also have to repent of the old harmful story.

I want to talk about one form of sin, racism and white supremacy, and how they manifest in climate racism. Racism is about institutional power, not personal prejudice, and we live in a world that gives more power and agency to white people.[13] It means that people who are white don't have to worry about being followed in a store, finding a Band-Aid that matches their skin color, being denied a job or a spot in university because of the color of their skin, or learning that their children have been shot just walking down the street or while driving a car. This system of racism is why people who are white usually have better access to health care and education and positions of power and decision-making. It's also why white folks weren't affected economically by the kind of redlining that Michael described.

It does not mean that individual white people are bad—or more sinful than other people. It does not mean that people who are white—people who are connected to historical and institutional power—have the intent to be racist. But even when we don't mean to be hurtful, pain still hurts. Racism is a sin precisely because racism breaks our connections to each other, God, and creation.

And our connections run deep if we let them. In Genesis 2, God forms the human out of earth and breathes life into us earthlings. The text says, "The Lord God formed the human from the topsoil of the fertile land and blew life's breath into his nostrils. The human came to life" (Genesis 2:7, Common English Bible).

This is a story that reminds us that we—all of us—are made in the image of God and connected to the ground. When we break connections, when we sin, we forget that we metaphorically come from the same ground—from the same ashes, dust, and topsoil, later to return back to the same ashes, dust, and topsoil.

Racism is a sin because it means we forget our common ground, it means we disconnect ourselves from others. The sin of racism infects other parts of our lives, including our ecological lives. People of color and people who are poor—populations that often but not always overlap—are hurt by environmental destruction first and worst. As you know from chapter 2, this is called environmental racism.

We live in a world whose climate is already changing. We have passed many of the tipping points in climate change that climate scientists have said will be irreversible in our lifetimes—atmospheric carbon dioxide levels are over four hundred parts per million, more than ever before.[14] Those already experiencing most of the effects of these changes live in the Global South and on islands populated by people of color, people who are least responsible for climate change. The people who have emitted the most carbon live in the Global North, communities that have yet to experience many of the effects of climate change and who sit at the powerful tables of climate policymaking.[15] Basically: Global North governments get to make the decisions and laws that keep pollution coming, reaping the benefits of fossil fuels through industrialization, without worrying about the impacts of those fossil-fueled decisions. *(This points to the fact that we need a fund for loss and damage that does not harm those in the Global South and other vulnerable countries. Those that do the least harm should not suffer under a green economy. Not only should Global North governments pay their fair share for the damage that it has caused, we should pay climate reparations. This cause is just and holy. —Michael.)*[16]

Climate change policy has been made in ways that don't recognize the humanity and human rights of other people, of people who are poor and people who are on the front line of climate change. *(And, white supremacy erases those who are Indigenous. Erasing Indigenous people silences their voices. By silencing the voices of those who are Indigenous to the land, you lose the lessons of stewardship for the land. There are far too many examples of this foolishness and the consequences that result. This is another reason why white supremacy makes no sense. —Michael)*

They have lost and will keep losing everything, like in the biblical story of Job. We know his story—of the deep and horrendous suffering and angst of a man who lost everything. His land is wiped out, his family is killed, and he loses his livelihood. He so desperately wants to know why this has happened to him. We often read the story and quickly skip to the part when his wife tells him to curl up and die, and his friends try to come up with all the reasons why this is happening to him—responses we might have to people who suffer from all sorts of issues, including climate change. We tell that story as if Job's loved ones want to skip to the part where things get better. But before they do anything, they come to be with Job. The text says they leave their homes and come and see their friend. His suffering has been so great that they do not recognize him.[17]

Their hearts break open and they weep for his plight. The word that gets translated as "console/condole" is a Hebrew word, and it means "to shake one's head—to be speechless in one's commiseration."[18] They throw dirt—a symbol of our connectedness and createdness—above their heads in a ritual act of aching despair. And they sit in solidarity with Job for seven days. They sit with him and wait for him to speak. They wait with him on the place that connects us, the ground on which we live and move and have our being. They wait in that time of confession until they feel the weight of just how much Job is hurting.

What will it take for *our* hearts to break open and for us to sit on the ground—the ground that connects all of us from our beginning? What would it take to sit with the suffering of creation? First we have to recognize and name those pieces—those sins of racism, environmental crisis, and climate—even though it makes us uncomfortable.

In Luke, on the road to Emmaus, the disciples are grieving and confused, so consumed by those feelings that they do not recognize Jesus. I am such a huge fan of the disciples in the Bible because they're impeccable symbols of who *we* are in the world. Of course they should recognize Jesus, just as of course we should recognize the dignity of other people and other parts of creation. They do not yet believe that Jesus the Christ is alive—they only know that the man they loved so dearly has been killed. They have been sitting on their own suffering ground, and they have just heard the impossible stories that the body of the man they knew and love is no longer in the ground.

In the shocked questions they pose to Jesus, can you feel their confusion and despair? This is how I feel when I spend too many days focused on the realities of climate change that disproportionately hit communities of color and then come across someone who asks me, "What are you talking about?"

Jesus explains to the disciples what they don't know—teaching them as always about the connections around them—and even though they don't recognize him, they ask him to stay with them. These texts from Job and Luke remind us of the great suffering, the deep groaning of creation, and our sadness when we feel the disconnection between us, God, and creation. They remind us of our need to stay in those moments of silence, recognizing and confessing our sins.

But then, because finally the disciples see Christ in his fullness—feeding them and loving them—they recognize him, and then they feel all their broken pieces being refit together. This is why I keep doing this ecological work through the lens of Christian faith. Christ's resurrection comes into a crumbling world, a world marked by the sin of racism and disfigured by climate change, a world in which people anguish because their life is not recognized, a world flooded with emissions that are killing us all—but the most vulnerable first.

I don't know if we can turn the tide on climate change. I don't know if we who are white will ever fully give up the sin of racism, including climate racism. These are sins that we must see, recognize, and confess. But I do know this: In Genesis, the ground we're made from is the topsoil[19]—the hearty two inches of earth from which life comes—and in the days and months and years after Easter, we receive again the breath of God in the resurrection. This breath resurrects and reconnects us. The ground is alive—just like Christ—and so, now, we can get up and get on the road to action. Then we can change the story.

Invitation for Action

There are so many stories of experiences of climate injustice and economic injustice linked to racism in America. In this chapter, Michael has shared some of his story as part of that larger story. It is embedded in a societal myth that somehow racism does not really exist in America, or at least that it is not so bad. This lie subtly says we are all equally vulnerable to suffering from climate change. We can act to change the story—to reveal the realities of racism and climate racism—so that, by changing the story, we open doors to resisting the wrong, changing the rules, moving the money, building the new, and growing the bigger we in the sacred struggle toward climate justice for all. One way you can take action is to learn more about climate justice and the stories of people who have been struggling for better access to clear air, clean water, clean land, and for energy policies and climate policies that are aligned with climate justice—people whose stories of action are changing the larger story. One place to explore is the Climate Reality Project's Climate Justice Resources: https://www.climaterealityproject.org/climate-justice.

How can you learn the stories of people already trying to change the story in the place where you live? How can you be part of those stories? To begin, look for organizations that already exist. For starters you might:

1. Look at any of the organizations we've mentioned in the past chapters.
2. Work with others to locate one or two organizations led by Black people, other people of color, or Indigenous people that are addressing climate change, environmental racism, local agriculture, or equitable access to renewable energy, food, or water.
3. Learn from that group, read their materials, listen to their stories, and find out how you or your faith community can be a part of their work.
4. Find specific ways to publicly amplify their story and their work to change the story, always checking first with the group to see whether your means of amplifying seems good to them.

To learn more and to meet a global community of companions, be sure to visit the website for this book series at https://www.buildingamoraleconomy.org/. Select "Book 1: Climate." There you will find a treasure trove of resources—organizations and websites, reading material, film, and more—to accompany you on the faithful journey toward climate justice.

Blessing

Put your hand on your belly and breathe.

Remember that you come from the topsoil of the planet.
Remember that your breath comes from God, who loved you and all that is into being.

May this remembering connect you to the struggle for climate justice.
All who breathe are made of topsoil and love.
May this remembering connect you to the stories of others.
All who breathe are made of topsoil and love.
May this remembering connect you to our ecological community.
All who breathe are made of topsoil and love.

Remember that you come from the topsoil of the planet.
Remember that your breath comes from God, who loved you and all that is into being.

CHAPTER EIGHT

Live Lightly

MUCH OF WHAT we have written so far has been about the systemic interactions between climate change, white supremacy, and capitalism. We've noted how we have to change the rules of how we live, change the stories that guide us, resist what is wrong, build a new reality, and rely on the bigger collective community. These fingers on our hands of action are important. In this chapter, we move into living lightly, which will focus on individual action and daily practices. These daily practices (in transportation, food, and energy use) contribute to more ecological, equitable, and democratic economic lives, communities, and society. These practices can shape who we are as practical and spiritual beings who then shape other parts of the world. But we have at least three caveats to this framing.

First, it's important that we write this chapter without trying to greenwash the situation. Greenwashing is a deceptive tactic that companies and governments use to pretend that they are further along in their emissions-cutting efforts than they really are or to hide systemic bad practices behind a single positive aspect of a product.[1] We are not going to sugarcoat the situation, and we aren't going to suggest that we can buy our way out of the climate crisis. We also do not mean to suggest that we have the answers. Greenwashing by companies and governments reminds us that we—communities of care—must be fiercely honest, and that, while we can't rely on these institutions to save us and the planet, we must hold them accountable to do their part and to be honest.

The practices and mindsets we reflect on in this chapter are not product-based, and we seek to explore how individual practices of living lightly on this planet we call home impact us and our commitments to systemic change.

Second, climate change is real and is caused by humans burning fossil fuels at a rate that is unsustainable and out of balance. And we know that emissions from just one hundred companies have contributed 70 percent of emissions that have caused climate change.[2] There is deep need to make changes in how each of us exist on the planet, and those changes need to be in relationship with systemic changes. While our individual or household changes alone are vitally important, they are not adequate. We can't buy or recycle our way out of climate change (though companies and the fossil fuel industry have tried). We can't and shouldn't think that buying compostable plates will save the rainforest or Indigenous communities on its own.

Third, individual changes are not always possible for everyone. Not everyone can switch to solar panels or go vegan or drive an electric car or take public transportation, for lots of reasons. Some of these reasons are economic, like paying for a solar panel to put on a building that you own. Some of these reasons are about access, like the ability to live in a neighborhood that has vegan food or charging stations available. Some of these reasons are linked to infrastructure, like how many places in the United States still don't have reliable, timely, and affordable public transportation. This caveat is often about how poor people can't afford things, let alone high-priced, environmentally sustainable things. This is not what we mean. Such an argument is too simple and ignores how people who are poor often have creative ways of stretching a dollar that are also sustainable (like going to a thrift store for clothes, which is just as ecologically committed as purchasing an organically and sustainably made new T-shirt).

In short, there are limits and nuances related to the impact of individual action, and people and the planet deserve systemic change that responds to the reality of climate change. So, in this chapter, we write about living lightly, but always in the context of the other fingers on the hands of change. This contextualization looks like remembering that we all need more than individual action; we need collective action and organized efforts to move those with political, economic, and social power to make systemic change, too. Michael wrote in the previous chapter that systems of white supremacy and racism are personal for him as a Black man in the United States. Without speaking for other Black people, he explored how his particular life story has been impacted by white supremacy and its economic implications. In very different ways, living lightly and making individual changes for an ecological life is deeply personal for abby, impacting her from childhood and her personal and professional life. (*One* of the differences here is that Michael can never escape white supremacy and abby is free to ignore her ecological vocation.) The rest of this chapter is written from abby's point of view, with Michael weighing in.

I am the way I am because of my parents. My mom, who read through this whole manuscript before it was published to make sure that we actually make sense, has a sticker on one of her reusable water bottles that says, "Leave it better than you found it." She told me recently, "that's my life's motto and you can quote me on that." In that same conversation, as we sat in my parents' living room, we were talking about how she and my dad picked up trash on the way back on their walk that morning. As we talked, I couldn't help but think about the yard behind us, visible through the big glass windows of their home. The yard is full of native plants. My mom claims she doesn't have a green thumb. Yet, here is an acre of arable land she has nurtured with plants picked for their support of pollinators, prairie flowers and grass, and native trees planted to restore some of the canopy.

Tucked into the corner of the yard is the compost pile. I often turn to the compost pile when I try to pinpoint why I have spent my life in religion and ecology. One of my core memories of my childhood is of my dad shoveling out the path from our house to our compost pile anytime it snowed. He still does this, and I do it for my parents when I'm home and it's snowed. It is an act that communicated to me that we would continue to use the compost pile even when it snowed, even when it would have been so much easier to not make the trek in the bitter cold. This small individual action has had systemic implications: It showed how in our family we value caring for the planet, even when it's inconvenient. That value shaped me (and my sisters) and how we show up in the world.

This shaping is, I think, the real value of individual action. What we do in our own corners of the world impacts those corners. It also impacts who we are and how we are in community. We take individual action to nurture ourselves so that we can also work on a systemic level. Now, some of us are so marginalized by the larger systems that to just survive is a struggle. Again, I am not here to shame those folks into living out ecological values in their own lives. I am not here to say that we have to shame poor folks into using reusable bags or wearing organic T-shirts. Should they have that option? Yes. Everyone deserves good, quality, reusable things. Is that going to save the planet on its own? No. Please go give the fossil fuel industry all your anger, thanks.

(I will say that we should listen to poor people. Poor people have been living sustainably out of necessity. Where I'm from, we call it survival. We did not go to thrift stores. We were given hand-me-downs. By devaluing the voice and experience of those who are poor, we lose a great deal of wisdom. —Michael)

When we were organizing for divestment from fossil fuels, many people resisted our call for a systemic action by saying that it felt hypocritical to ask the Presbyterian Church (USA) to divest from fossil fuels when they themselves were driving a gas-powered car or using oil and gas to heat their homes. In fact, one of the reasons that a group of us walked to the PC(USA) General Assembly in 2018 is because so many people told us that they thought it was disingenuous that we were asking for divestment from fossil fuels at an event to which we had flown. I wanted to change the conversation so that, when someone said, "Who are you to ask us to divest? You got here by plane, didn't you?", I could say, "No. I walked." (I also got my orneriness from my parents. Hi, Mom and Dad!)

Two things happened with that individual decision in response to a systemic problem. First, the group of us who walked over two hundred miles to that General Assembly spent two weeks walking, praying, and slowing down. We got to see the Midwest up close, and we got to connect with each other and how our faith in God sustains us in the work. We had time, because we were walking, to talk with each other about how God had called us to the work of loving the planet and other people. We got to see wildlife that we would otherwise have been whizzing by (I still get texts about the big turtle we saw on the side of the road!). It was incredible, and the slow pace set us up to be more grounded and clear-hearted at General Assembly.

Second, we got very clear about how the impact of our walking, instead of driving or flying, was minuscule on our carbon footprint and on the emissions of the world.[3] We did need way fewer fossil fuels to walk than if we had driven or flown. But we also needed a couple of support vehicles, and most of us flew home. It also took many hours of planning and fundraising. It wasn't a sustainable project, something any of us could take on for future long-distance journeys.

Daily Practices

Transportation

Transportation is one of the biggest personal emitters of fossil fuels. According to the Second United Nations Global Sustainable Transport Conference in 2021, "ninety-five percent of the world's transport energy comes from fossil fuels."[4] Much of that comes from cars and planes. We need to rely less on individual transportation (like cars) and more on communal transportation (like buses and trains). We need to use more public transit if we are going to stop the growth of climate change.[5] But we lack infrastructure in many parts of the world and in the US in particular. We need, according to the World Resources Institute, to build a lot more infrastructure pretty quickly: "Cities need to grow their rapid transit networks six times faster by 2030 to shift enough trips away from private vehicle travel to align the transport sector with 1.5-degree-C climate goals. In the world's 50 highest-emitting cities, rapid transit needs to double between 2020 and 2030."[6]

We need not only more infrastructure but also more equitable infrastructure. Where I live on the North Side of Chicago, it's easy to jump on the train to get almost everywhere I want to go in the city in a reasonable amount of time. But if I lived on the South Side of Chicago, it would be harder to access trains, and buses would be more irregular and slower. TransitCenter notes that "pervasive racism and discrimination in land use, transportation, and transit planning have resulted in wide gaps in transit access across race, income, and other characteristics, worsening social inequity."[7] Essentially, in Chicago, you're more likely to live near reliable, affordable, accessible public transportation if you're white and have a stable income. And this is just in Chicago. If I want to get to my parents' house, I have two bus options—and a train option if I take the Red Line to a bus, and then a regional train, and then my parents still have to come get me from a station about forty-five minutes from their house. It's doable, but it's more work. You know what's easier? Getting in my car and driving.

Still, recently my child and I took the bus to my parents. We talked to the other people waiting to get onto the regional bus, and then we settled in our seats. I got to actually talk to my child while we went on our way, and she had a much better view of the cornfields I love.[8] It was a minuscule change in our carbon footprint. It was a big shift in our relationship, in our time together on that journey. I think that's mostly what living lightly is about. It's not about how we change the systems; it's about how we change ourselves and our closest relationships.

Food Choices

Another core memory from my young adulthood is of going to a conference with my college chaplain. I don't remember what the conference was. I do remember the two of us talking about how, if I was going to do ecological work for the rest of my life, I had better become vegan (rather than just a vegetarian) in order to be taken seriously in the movement. I tried. I also really like cheese and ice cream a lot, so I'm now just a vegetarian. And it's true: Eating lower on the food chain lightens our impact on the climate, especially since so much of normalized, affordable meat relies on industrial agriculture that pillages that planet.[9] When I was working for GreenFaith and needed to travel around the globe to organize with other people of faith, one of the commitments I made to myself was that I would be vegan on those trips (as much as possible, and with the caveat that I would be a good guest to hosts, especially in the Global South). This commitment was about recognizing that I was emitting a lot of fossil fuels in order to travel, so I needed to find other ways to scale down. But it wasn't about saying "I'm saving the world by being vegan." It also wasn't about saying "Look at me—I'm a better environmentalist because I'm eating lower on the food chain."[10] Instead, it was about doing the internal work of finding alignment and balance in my actions. It was this alignment and balance that made me feel like I could join other people at the table. I felt grounded and connected to my sense of call to love God's creation. (Sort of how like when we have invited you into breath prayers at the end of several of the previous chapters. Breathing keeps you alive, but it doesn't change anything beyond you. It's so important and it's deeply personal, in that it is a process for your body alone.) The individual action felt like a spiritual practice, an attention to my body and God's body that prepared me for the work ahead.

Energy Use

Finally, early in my time in the movement, I worked with Faith in Place on a weatherizing campaign. In the next chapter, Michael notes the weatherizing campaigns he's led as part of the People's Justice Council. Weatherization is a process that seals up a dwelling so energy isn't wasted.[11] We passed out supplies to seal windows with plastic and provided information on and support for covering pipes and sealing the bottom of doors. It reminded me of light bulb campaigns, in which congregations and communities encouraged people to swap out incandescent light bulbs for more efficient LEDs. These are small steps that impact the planet while also making energy bills lower. I still have to weatherize parts of my home, and I still wonder why and how we make poor people live in spaces that aren't tight by original design.

Lowering our energy use is one more way we can lighten our impact on the planet. It's a practical step, and the process of switching out light bulbs, windows, HVACs, and other systems can feel like we're having an impact—it can feel really good to see the difference in energy costs and energy consumption. So much of our impact can't be measured, and this is one way we can celebrate a win.[12] When I was directing the Green Seminary Initiative, we asked participating seminaries to do an energy audit. How much were their electricity bills? How much were their heating and cooling bills? How were lights beings used? And then what were the concrete steps they would take to reduce energy usage and the related bills? Some of this meant weatherization and switching out bulbs. And for some schools, like Pacific Lutheran Theological Seminary, it was a moot point because they sold their campus and moved into a single floor in a shared building. They shrank their footprint to a small percentage of what it had been! Now, that move was motivated by lots of circumstances, and part of it was ecological. Their energy use plummeted, and they got a beautiful and consolidated space that is perfect for their present reality. (Their classrooms and offices are built out around the chapel space, which is at the center of the whole floor. They have literally centered worship in their space.) These practices ask us to think about the kind of energy we are creating while also lowering our bills.

Still, until we invest in making renewable energy affordable for impoverished people and in different infrastructure, like ordinances that require buildings to be tight or utilities to be solar or LED by default, these practices are individual, with small systemic impact. That is why it is so important to link living lightly with changing the rules, building the new, moving the money, and even changing the story about how we are to live our daily lives and what matters in life!

I think about the annual practice of weatherizing our home. It's a ritual of preparing our home for winter. It shapes how we will act for a season, sort of like a season in the liturgical calendar. We are changed for a little while by steps that impact our energy use. We are changed—and prepared and supported as we also seek to build the beloved community.

And as I think about the beloved community, I wonder about the early church in Acts. I can't help but think about how if they were creating church now, they would be organizing car pools and vegetarian/local meat potlucks. They would be the first to volunteer for weatherization programs, armed with sealant for windows and doors as they went door to door around their communities. Maybe they would have LED light bulbs tucked into their pockets, bulbs to pass out to their neighbors. I imagine them getting their food from small local and often urban farms (community supported agriculture, or CSAs). Each of these practices would build community awareness and put small measurable changes in actions. *(Amen and Ashe, Doc. —Michael)*

These actions have to be contextualized into the larger structures of our world. We cannot blame individuals, and especially people who are marginalized by corporations and governments, for not saving the world, for not stopping climate change. Living lightly is just one finger on the hands of change. We need each finger, on both hands, engaged in this work to build a moral economy for all.

Invitation for Action

Beyond Plastics is an organization started in 2019 to end plastic pollution as part of a larger struggle against the fossil fuel industry. They talk about how "it will take changes at every level of our economy and civil life to stem the tide of plastic pollution. Individuals need to be moved to act in their personal lives and take action as part of a growing movement; corporations need to feel the pressure to initiate changes in their purchasing and packaging habits; governments need to impose bans and adopt laws that require extended producer responsibility; and new manufacturing of plastic has to be prevented from spreading."[13]

As we have said, individual action alone will not save us. Beyond Plastics has built out a toolkit for congregations and people of faith to use as they change their personal and communal plastic use *and* advocate for different laws around plastic manufacturing. Find more on their website at https://tinyurl.com/FCBPTK. Look at their toolkit and explore how your congregation, faith community, or municipal community can take on using less plastic *and* advocating for laws that lessen plastic production and use as part of a moral economy and as a step on the path toward climate justice.

To learn more and to meet a global community of companions, be sure to visit the website for this book series at https://www.buildingamoraleconomy.org/. Select "Book 1: Climate." There you will find a treasure trove of resources—organizations and websites, reading material, film, and more—to accompany you on the faithful journey toward climate justice.

Blessing

Remember that you were made from the topsoil of the earth and the breath of God.

Remember that all that is created is made from that same topsoil and breath.

Remember that we belong to each other.

Remember that who we are, what do we, how we are is rooted in God's love for us and God's desire that we care for each other.

Remember that we, small that we are, are not the systemic problem. We do not need to carry the full load of guilt. We can hold corporations and governments accountable.

And:
Working together as a "bigger we" we can move them to change. Remember that together, little by little, we can change the world.

CHAPTER NINE

Listen and Amplify

THROUGHOUT THIS BOOK, we've brought in stories of people who are working to change the rules of the economy and community. This work creates a world that we know is possible. These stories help us believe that we really can create a new story, we really can resist what is wrong in the world, we really can live together more lightly. In this chapter we continue to focus on stories. We seek to listen more deeply and then to amplify the stories and experiences of people who are doing the work, who are living and organizing at the intersections of white supremacy, capitalism, and climate change. We—all of us, but especially those of us who aren't marginalized by systems of oppression—have to find postures of listening. Then we must find ways to amplify these stories and to let the work of others impact our lives. In this chapter, Michael leads us into listening and seeking stories in our own communities that we can amplify.

abby and I attended a leadership retreat in Maine during the summer of 2023. I can remember it getting insanely cold during the night. I felt completely lost because I'm used to the South. It is burning up 24-7 in June in the South. I remember thinking to myself, "White people wear sandals anywhere." *(This is data-driven. We will wear sandals everywhere. —abby)* As we enjoyed our walks among and communing with the land and nature, I remember asking about those whom nature has turned against. How do they view nature? Who's speaking for them?

The Call to Listen . . . and Share the Stories

I am the founder and executive director of the People's Justice Council. Our mission is to engage and equip communities with tools and access to build power from the grass roots up for change at the policy level ("changing the rules"). We have had the honor of engaging communities throughout the southeast and Gulf South. When I reflect on listening and amplifying, I think of the communities and the people I have encountered.

Such communities include the one we were asked to engage with in South Fulton County, Georgia. This predominately Black middle-class community, with its multilevel homes, was under assault because of an illegal landfill that was on fire. This landfill was placed in this community without any feedback from the community. It has operated illegally for years, and it has been protected by the local government. This middle-class community is suffering from the same environmental injustice that poor Black, Brown, and Indigenous communities suffer from. *(This is such a good example of how it is racism, not just classism, that makes industries and governments turn communities of color into the locations of environmental degradation. —abby)*

One program of the People's Justice Council is Moving from Resilience to Restoration. In 2023, we were blessed to tour the Gulf South to engage with our network of program hub partners. One hub partner, Mississippi Rising, is not only providing safe, drinkable water throughout Jackson, Mississippi, but also providing fresh food in communities suffering from food apartheid. In 2023, we visited with directors Lea Campbell and Morris Mock.

During our conversation, Morris said, "It's apartheid. When you look at the poisoning of water, the lack of food sources, and lack of access, what else can it be?" We traveled to Eupora, where we met with Sheena and other community members at the Fannie Lou Hamer Center for Change. During our discussion they lamented over the lack of assistance in times of disaster. I observed that there were a lot of churches in the area. I asked about the church's participation. Their reply was that the churches did not want to help. They saw it as a political action and did not want to be involved in politics. These people felt abandoned. Yet, when an unnatural weather disaster happens, the Fannie Lou Hamer Center for Change is right there, providing disaster recovery assistance, despite scarce resources and lack of assistance from elsewhere. They do it anyway.

While in Jackson, we met with Esther and members of the Unitarian Universalist Church of Jackson. The sentiments of apartheid and abandonment were shared on this visit also. The Unitarian Universalist Church of Jackson was gifted an old bottling plant with an artesian well. The well is certified and can provide water to communities throughout Jackson. The well is not operational because it needs a pump and maintenance. The cost is about $25,000. Because of where they are located, they have not been able to get the assistance to make the well operational, though now they are working with Mississippi Rising Coalition to fix the well and support clean drinking water for their community. But, when the surrounding communities are in need, the Unitarian Universalist Church of Jackson steps up to serve their community. While they might not have the funds, they do it anyway.

Through the People's Justice Council's Weatherize Every Residence in the Southeast (WERiSE), we teach community members how to weatherize their homes. We believe that the process of weatherization can be a shared community activity. Think of a community cleanup day that centers weatherizing homes. This is also a great jobs training and creation program. Since implementing this program in 2019, we have held several workshops and weatherized several homes. In 2024, we received our first significant grant for this work. Prior to this, we had been doing this work through in-kind donations. We did it anyway.

I provide these and other examples to show what communities sitting on the outer edges of privilege must deal with. We are often blamed for conditions that were created for us. We are isolated from resources. We are underserved. We are underpaid. Yet, we are surviving. We are surviving because of the people and organizations in our communities who center care. Many of these people and organizations are often under-resourced and overlooked.

What would it look like if we no longer had to "do it anyway"? How much impact could we make if we didn't have to share testimony that the churches were not helping us for their varied reasons, and that government agencies and funders were ignoring our people and their needs? Why aren't these stories well-known?

We were called to Newbern, Alabama, to meet with Mayor Patrick Braxton in 2021. He was the first Black man to be elected mayor of Newbern, and we were the first to report on his story. We were called because he was unable to take office and was locked out of his office by the white former mayor and his family. Not until Mayor Braxton filed a civil rights lawsuit against Alabama was this nationally publicized. We (the People's Justice Council) were never given credit for being the first to amplify the story. Another insult was that the images that we took were utilized and we were not given credit for them.

The writer of Acts understood that it was the Holy Spirit who empowered the early Christians to care for one another. It was caring that drove them to sell all that they had so that none would be without. However, how did they know what to care for? They knew what to do and who to care for because they listened to the stories of needs rather than ignoring them. The stories were told and heard.

In our travels, we found many leaders and organizations that are having major impacts. Equitable mutual aid and disaster recovery relief are not happening in Jackson and Eupora alone. Mutual aid, disaster recovery, sustainable affordable building, co-op farming initiatives and the likes are happening throughout the south and beyond.

The Black Hive is the environmental and climate justice table for the Movement for Black Lives. We are a cohort of Black climate and environmental justice experts led by Natalie Jeffers.[1] We use our collective experience and knowledge to assess how climate change and ecological destruction impact Black communities in the US and across the global Black diaspora. Together, we organize for cleaner, better, and safer futures for Black lives.

It was through the Black Hive that I met Dr. Marsha Mitchell-Jackson. Dr. Jackson is the director of Southern Sector Rising. Through the work of Dr. Marsha, I learned about Sand Branch, Texas. Sand Branch sits fifteen miles outside of Dallas. It is a poor Black community with no running water. Yes, in the United States of America, there are communities with no running water. Southern Sector Rising and the Chisholm Legacy Project are providing solar water tanks for homes in the Sand Branch community.

Leaders, organizations, and communities such as those just mentioned need to be amplified so that their realities are far more widely known. What better way for communities of faith to get involved than through listening and then amplifying? The book of Isaiah asks, "Watchman, what of the night?" (Isaiah 21:11). This was a desperate cry for help that asked the watcher how much of the night was left. In essence, it meant "How much longer will it last?"

Communities of faith are seen as the watchers who are being asked "How much longer?" As communities of faith, we are obligated to answer the question. The only way communities of faith will know how to answer is if they are listening to the communities that are suffering and the leaders and organizations that are leading with real solutions. By listening, the People's Justice Council has developed platforms for advocacy that prioritize what communities need.

Invitation for Action

Often we see a problem and seek to fix it in big ways. I commend big initiatives when they are grounded in the leadership and voices of the people who are suffering, initiatives like the Fossil Fuel Non-Proliferation Treaty Initiative, which demands a legally binding, international treaty called for by frontline communities in the Pacific Islands. However, many solutions are also happening at the grassroots level. Networks such as the Southeast Climate and Energy Network, Climate Justice Alliance, US Climate Action Network, Green the Church, GreenFaith, Interfaith Power and Light, the Black Hive, Climate Critical Earth, and other grassroots networks are offering powerful solutions and need the church to listen, amplify, and support them in other ways. Organizations such as People's Justice Council, GASP, Sol Nation, New Alpha CDC, the Imani Group, and other grassroots organizations need the support of faith communities, including their respectfully listening ears and their ability to amplify into the broader public. Leaders need support too. Leaders like Esther Calhoun, Vernice Miller-Travis, Kari Fulton, Rev. Dallas Conyers, Rev. Laura Kigweba, Rev. Dr. Gerald Durley, Dr. Mildred McClain, and other grassroots leaders need amplification. We need our stories told, and people need to hear so that they will begin to see and take action.

Our actions must move beyond thoughts, prayers, blessings, and well wishes. Just as these networks, organizations, and leaders are making their impact tangible, so should you, the reader. You could find a group that is working on a problem related to climate change, racism, or economic injustice that really concerns you and listen closely and respectfully to what its members are saying—both about the problem and about solutions. We have shared several examples. We invite you to visit their websites and learn more.

Then, find ways to amplify their voices, to bring their stories into more public light. With their guidance, can you publish a series about them in your faith community or other group's newsletter or social media channels? Can you invite them to speak (offering an honorarium) at a school or other public venue? What other avenues do you have for bringing their perspectives to the broader public, always doing so in ways that they affirm? Then, go on to invest. Invest in tangible ways. Invest with your time, your talent, and your treasure. I challenge each of you to find a network, organization, or leader that is working on an issue you are interested in, listen to them, and give them your full support.

As you find places and people to support, there will be so many moments when it will feel hard to listen. It will feel difficult or uncomfortable to amplify stories. Do it anyway.

To learn more and to meet a global community of companions, be sure to visit the website for this book series at https://www.buildingamoraleconomy.org/. Select "Book 1: Climate." There you will find a treasure trove of resources—organizations and websites, reading material, film, and more—to accompany you on the faithful journey toward climate justice.

Blessing

The Challenge

He has shown you, O mortal, what is good. And what does the Lord require of you? To act justly and to love mercy and to walk humbly with your God.

—Micah 6:8

The Blessing

The LORD bless you
and keep you;
the LORD make his face shine on you
and be gracious to you;
the LORD turn his face toward you
and give you peace.

—Numbers 6:24–26

CHAPTER TEN

Practice Awe, Gratitude, Lament, Holy Anger

HERE AT THE end of the book, we imagine that you are feeling many different feelings. If you made it this far, you've encountered how Christianity influenced the beginning of the fossil fuel industry, how climate change is getting worse each year, and how people of color and people who are poor experience climate change more harshly than groups who are not marginalized because of racism and classism. You've also encountered how people are resisting, creating new rules and new stories, how our faith stories can provide examples of people embracing community instead of empire. There are so many ways to engage all the feelings that emerge in the face of climate injustice, especially in the knowledge that economy and faith are intertwined.

In the preceding pages, we've woven gratitude for others and the planet, awe at the beauty and goodness present on this earth, lament for all that has been lost, holy anger for all that has been done—all of this flowing into the Invitations to Action and the Blessings for each chapter. These feelings are important. We have to allow ourselves to experience, share, and be moved by awe at the miracle of life, lament for the unnecessary suffering and all that has been lost, anger at what thwarts fullness of life, and gratitude for all that is good, life-giving, liberative, and healing. Nothing is linear, not in destruction and not in healing. This—practicing awe, gratitude, lament, and holy anger—is the ninth of ten fingers on the hands of action that we've walked through together.

There is a tenth. In this book series, we call it "Drink the Spirit's Courage." For people of faith, this means drawing upon the practices, stories, faith claims, wisdom, and other resource of our faith tradition to feed our courage and one another's courage for the journey ahead. The pages of this book overflow with examples, from Ruth to the early Christians to people at work in the world today. These—faith forebearers and contemporary companions on the journey—are nourished by what Christians know as the Spirit of God. People in other traditions know this Spirit by other names.

The Spirit nurtures courage and hope in other ways, too, beyond what we think of as religion. For us, one way is human relationships. You have heard our letters to our children. Our relationships with them and with other loved ones, trusted companions, and people who challenge us and help us to grow all nourish our courage for the journey toward climate justice. Many people draw courage and hope from the earth itself, its waters, creatures, lands, and growing things. One friend tells the story of leaving her shelter filled with despair, only to find her heart brought back to life by a lone bird singing with abandon. Art, too—music, dance, poetry, painting, street art, and more—inspires and sustains courage for the work of justice. We hear the tambourine of Miriam accompanying the Hebrew people in their escape from Pharaoh, the psalmists' harp and song, the angelic voice of Billie Holiday singing about "strange fruit," the poets quickening the hearts of liberation struggles in the Caribbean, the heroic reporting of Ida B. Wells-Barnett as she wrote about Black people being lynched, the street murals of Chileans resisting a dictator, the resistance that existed in the hush harbors, the sounds of the Black church leading the first demonstration against environmental injustice in Warren County, North Carolina, and so many more artists throughout the continents and ages.

Yes, religious faith and its practices, relationships, the earth, and art all are vessels of the Spirit's courage. For you, how does the Spirit nourish your courage and hope? How could you open yourself ever more fully to drink from those vessels? How might you do this with others, in community? This is the tenth finger on the hands of action—the hands of healing change.

As you now review the ten fingers, how do you see their connections? Which do you want to dig into more deeply? With what people? Which are in your wheelhouse? Which are stranger to you and might challenge you to grow?

Build the New—creating or supporting new life-giving financial systems, businesses, and agriculture
Change the Rules—changing public policy and laws
Move the Money—moving to a triple bottom line, divesting and reinvesting
Grow the Bigger We—building alliances to heal exploitative economies
Resist the Wrong—saying no to extractive and exploitative economies
Change the Story—seeking a new story of what can be
Live Lightly—changing individual action as part of the larger whole
Listen and Amplify—paying attention to the stories of people who are on the frontlines
Practice Gratitude, Lament, Holy Anger—allowing ourselves to feel all emotions
Drink the Spirit's Courage—practicing the rituals and resources of our faiths and heeding relationships, the earth, art, and more to build a more equitable and humane world

Invitation for Action

A Time for the Spirit's Courage

Environmentalist David Orr writes that "hope is a verb with its sleeves rolled up." We each get asked where we see hope in the movement for climate justice. And to be honest, it's easy to lose hope when the fossil fuel industry has so much money, the United States government—the people who are supported by voters around the country—continues to support policies and politicians that extract and exploit the planet and other people, and international and local movements for climate justice seem to have little immediate impact. Yet we hear the New Testament's admonition and invitation: "Be prepared to give an account of the hope that is in you" (1 Pet. 3:15).

Throughout this book, we have tried to point to climate hope through the stories of people of faith working for a better world. These are some of the stories that bring us hope. We find hope in the struggle, hope in the fight for what is good. We find hope in the communities that keep doing the work of building faithful lives in relationship to each other, the planet, and God. Hope in the communities that keep showing up, taking action, and building the new. Is this enough? Where do you find hope?

And whether or not you find hope, we trust you to act. Act even if you fear you will fail. Act even if you're afraid that what you will do will be imperfect. Take courage, knowing that there are millions of people around the world taking action with you. Let the Spirit feed your courage to act, and let your actions feed hope.

Blessings

The final blessing for you are words of wisdom from a poet. May you be blessed on this journey toward climate justice and climate hope. May you remember that we belong to each other.

A Small Needful Fact

Is that Eric Garner worked
for some time for the Parks and Rec.
Horticultural Department, which means,
perhaps, that with his very large hands,
perhaps, in all likelihood,
he put gently into the earth
some plants which, most likely,

some of them, in all likelihood,
continue to grow, continue
to do what such plants do, like house
and feed small and necessary creatures,
like being pleasant to touch and smell,
like converting sunlight
into food, like making it easier
for us to breathe.

—Ross Gay

NOTES

INTRODUCTION

1 The story of Ruth and Boaz is in the book of Ruth, a short book in the Writings (*Ketuvim*) section of the Hebrew Bible and what Christians refer to as the Old Testament or First Testament of the Christian Bible.
2 The book of Acts is a book in the New Testament or the Second Testament of the Christian Bible.

CHAPTER ONE: WHO ARE WE AND WHO ARE YOU?

1 Matthew 6:21, New International Version.
2 Frank Trentmann, "Consumer Boycotts in Modern History: States, Moral Boundaries, and Political Action," in *Boycotts Past and Present: From the American Revolution to the Campaign to Boycott Israel*, ed. David Feldman (Palgrave Macmillan, 2019), 30.
3 One example of this is anytime someone tells you that folks won't show up to provide food or childcare when you need it, like after surgery.
4 Some of these examples include Lake Charles, New Orleans, Shreveport, Birmingham, Mobile, Union Town, Orlando, Panama City, Houston, and Port Arthur.
5 We also recognize that capitalism has taken different forms in different times and places. This book is not the place to examine those distinctions. We are talking, for the most part, about advanced global capitalism and its manifestations in the contexts about which we write.
6 Here we want to be clear that we are *not* saying that new economies for all should be rooted in biblical values. That would evoke a dangerous Christian nationalism. Instead, we mean to say that for us—for those of us who are Christians—we need to work for a world economy that values and supports beloved community because our Christian faith calls us to that.
7 For more about Ruth, explore André LaCocque's *Ruth: A Continental Commentary* (Fortress Press, 2004).
8 Cynthia Moe-Lobeda, *Building a Moral Economy: Pathways for People of Courage* (Fortress Press, 2024). https://www.buildingamoraleconomy.org/.

CHAPTER TWO: BUILD THE NEW

1 See Cynthia Moe-Lobeda, *Building a Moral Economy: Pathways for People of Courage* (Fortress Press, 2024.)

2 James B. Martin-Schramm. *Climate Justice: Ethics, Energy, and Public Policy*. (Fortress Press, 2010), 12. A synthesis of that groundbreaking report is available here: Larry Bernstein et al., "Climate Change 2007: Synthesis Report," Intergovernmental Panel on Climate Change Plenary XXVII. November 12–17, 2007, https://www.ipcc.ch/pdf/assessment-report/ar4/syr/ar4_syr.pdf.

3 Martin-Schramm, 128.

4 William E. Connolly, *Facing the Planetary: Entangled Humanism and the Politics of Swarming* (Duke University Press, 2017), 9.

5 Larry Rasmussen, "Eco-Justice: Church and Community Together," in *Earth Habitat*, ed. Dieter Hessel and Larry Rasmussen (Fortress Press, 2001), 1.

6 Throughout this chapter, I will work with the assumption that racism is prejudice plus power and that it is an institutional and systemic issue. The system of racism diminishes a person by defining them as just their race; a faithful response to climate change requires a person to come to the work as an assemblage of all their characteristics.

7 James Cone, "Whose Earth Is It Anyway?" in *Earth Habitat*, ed. Dieter Hessel and Larry Rasmussen (Fortress Press, 2001), 23–25.

8 Carol Wayne White, "Religious Naturalism: A New Ideal in Africana Religious Thought" (paper presented at the Transdisciplinary Theological Colloquium at Drew University in Madison, New Jersey, March 24–26, 2017).

9 Cone, 27.

10 Renee Skelton and Vernice Miller, "The Environmental Justice Movement," Natural Resources Defense Council, last modified March 17, 2016, https://www.nrdc.org/stories/environmental-justice-movement.

11 We promised you straight talk in this book. See Oyinkansola Ogunyinka, "Think Before You Take: 6 AAVE Terms Brands Should Use More Responsibly," Writer's Room, accessed February 1, 2025, https://writer.com/blog/aave-terms/. AAVE refers to African American Vernacular English. abby's twenty-year-old friend Charlotte Becherer assures her that this is way of speaking is also well-known and understood in her generation.

12 Cone, 27.

13 Amy Darymple, "Pipeline Route Plan First Called for Crossing North of Bismarck," *The Bismarck Tribune*, last modified August 18, 2016, http://bismarcktribune.com/news/state-and-regional/pipeline-route-plan-first-called-for-crossing-north-of-bismarck/article_64d053e48–a1a-5198-a1dd-498d386c933c.html.

14 Martin-Schramm, 7.

15 John Eligon, "A Question of Environmental Racism in Flint," *New York Times*, published January 21, 2016, https://www.nytimes.com/2016/01/22/us/a-question-of-environmental-racism-in-flint.html?_r=0.

16 Two forthcoming books in the *Building a Moral Economy* series will focus on food justice and water justice.

17 Kathleen Toner, "By Nourishing Plants, You're Nourishing Community," *CNN*, last modified on March 21, 2016, http://www.cnn.com/2015/09/24/us/cnn-heroes-joyner/. Joyner and his community were one of the subjects of a three-year-long research project on communities cultivating an ecological faith that I facilitated with three other people, funded by the Louisville Institute.

18 Pablo Martínez de Anguita, *Environmental Solidarity: How Religions Can Sustain Sustainability* (Routledge, 2012), 12.

19 Kathleen Tierney, "Foreshadowing Katrina: Recent Sociological Contributions to Vulnerability Science," *Contemporary Sociology* 35, no. 3 (May 2006): 207.

20 Cone, 27.

21 Wangari Maathai, *Replenishing the Earth: Spiritual Values for Healing Ourselves and the World* (Double Day, 2010), 149.

22 Cone, 25.

23 David Atkinson, *Renewing the Face of the Earth: A Theological and Pastoral Response to Climate Change* (Canterbury Press, 2008), 23.

24 Klein, 310.

25 Bernstein, "Climate Change 2007: Synthesis Report."

26 Jerry Melillo, Terese (T.C.) Richmond, and Gary W. Yohe. *Highlights of Climate Change Impacts in the United States: The Third National Climate Assessment* (U.S. Global Change Research Program, 2014), 48.

27 Klein, 417.

28 Klein, 409. Another example of the inequity of climate vulnerability: In 2006, the Food and Agriculture Organization of the United States released a report called "Gender: The Missing Component of the Response to Climate Change." At the end of the report, where they list the vulnerabilities and capacities of areas to respond to the effects of climate change, the authors do not even include North America and Europe.

29 Klein, 310.

30 Olivia Lang. "Maldives Leader in Climate Change Stunt." *BBC*, published October 17, 2009, http://news.bbc.co.uk/2/hi/south_asia/8312320.stm.

31 In the following discussion of sin, I will sit firmly in the Reformed Christian tradition as a white person in the United States. As such my exploration of sin is from the perspective of someone with a myriad of systemic privilege, which I will identify again below. Reformed Christianity is a branch of Protestantism. It includes Presbyterians but also some Anglican, Congregational, and Baptist communities. It is also sometimes called Calvinism.

32 Margit Ernst-Habib, "A Conversation with Twentieth Century Confessions," in *Conversations with the Confessions*, ed. Joseph D. Small (Geneva Press, 2005), 81.

33 White, "Religious Naturalism."

34 Shirley Guthrie, *Christian Doctrine* (Westminster John Knox Press, 1994), 22.

35 White, "Religious Naturalism."

36 David Atkinson, *Renewing the Face of the Earth: A Theological and Pastoral Response to Climate Change*. Canterbury Press, 2008, 44.

37 Guthrie, 219.

38 Guthrie, 215.
39 Daniel L. Migliore, *Faith Seeking Understanding: An Introduction to Christian Theology* (Eerdmans, 1991), 130.
40 The Hebrew (the original language in Genesis) word for human is *Adam*, and the Hebrew word for soil is *Adamah*. In the story of creation, God takes the soil (*Adamah*) and forms the human (*Adam*). Theodore Hiebert, *The Yahwist Landscape* (Oxford University Press, 1996), 6.
41 Karen Baker-Fletcher, "To Live as People of Dust and Spirit," in *Earth and Word*, ed. David Rhoads (New York City: Continuum, 2007), 12.
42 That is to say, our bodies are 60 percent water. Liz Cunningham, *Ocean Country* (North Atlantic Books, 2015), 22.
43 Whitney A. Bauman, "What's Left (Out) of the Lynn White Narrative?" in *Religion and Ecological Crisis: The "Lynn White Thesis" at Fifty*, ed. Todd LeVasseur and Anna Peterson (Routledge, 2017), 166.
44 I'm not yet clear about whether I think we can save ourselves in a theological sense or if I am too tied to the idea in Reformed theology that God's grace alone saves us. However, I'm also deeply committed to the Reformed conviction that by that same grace we respond with the best selves we possibly can. What insights might emerge if we consider how countries not responsible for climate change suffer the effects in relationship to an innocent Christ, who suffered on a cross for pushing back on unjust structures?
45 By "people with privilege," I mean people who benefit from global white supremacy and who contribute to climate change but are less vulnerable to devastation by it.
46 "Who We Are," Wisdom Keepers Delegation, accessed January 4, 2025, https://www.awisdomkeepersdelegation.org/.
47 Bauman, 170.
48 Maathai, 170.
49 Martínez de Anguita, 6.
50 For a deep dive into how Christianity supported the beginning of the fossil fuel industry, see Darren Dochuk's *Anointed with Oil: How Christianity and Crude Made Modern America* (Basic Books, 2019).
51 Kelsey Miller, "The Triple Bottom Line: The Triple Bottom Line: What It Is and Why It's Important," *Business Insights*, accessed December 8, 2020, https://online.hbs.edu/blog/post/what-is-the-triple-bottom-line.
52 See, for example, Oikocredit (https://www.oikocredit.org/), originated by the World Council of Churches. Later in this section we mention three other examples: one financial institution, one faith-based racial justice community, and one entrepreneur. We point also to mutual aid communities like the PO Box Collective in Chicago. There are resources like the Mutual Aid Map for Chicago https://www.communitykitchenchicago.org/mutual-aid-map) and the Mutual Aid Hub (https://www.mutualaidhub.org/). Mutual aid is a voluntary way for communities to support each other. In the next chapter, we'll explore more about how the early church was a bit like a mutual aid community.
53 Terra Rowe, *Toward a Better Worldliness: Ecology, Economy, and the Protestant Tradition* (Fortress Press, 2017) 2.

54 Ulrich Duchrow, *Alternatives to Global Capitalism: Drawn from Biblical History, Designed for Political Action* (Kairos Europa, 1997), 220.
55 Kevin Kruse, *One Nation Under God: How Corporate America Invented Christian America*, Basic Books.
56 "Our History," Earth Equity Advisors, accessed January 4, 2025, https://www.earthequityadvisors.com/team/#toggle-id-5.
57 In full disclosure, abby now uses Earth Equity Advisors, the company that Krull and Company became. Neither abby nor Earth Equity has received any compensation for its inclusion here.
58 "Frequently Asked Questions," Earth Equity Advisors, accessed January 4, 2025, https://www.earthequityadvisors.com/faqs/#toggle-id-4.
59 See for example the Solidarity Economy Network at https://ussen.org/, the members of the New Economy Coalition at https://neweconomy.net/member-directory/, the Worker Driven Social Responsibility Network at https://wsr-network.org/, and its Fair Food Program at https://wsr-network.org/success-stories/fair-food-program/.
60 Liz Case, "In the Beginning: How Arrayed Got Started," accessed February 1, 2025, https://www.shoparrayed.com/blogs/news/how-arrayed-got-started.
61 Koinonia Farm, "Who We Are," accessed January 4, 2025, https://www.koinoniafarm.org/about-us.
62 Koinonia Farm, "Come Stay Awhile and Serve," accessed January 4, 2025, https://www.koinoniafarm.org/come-stay-awhile-and-serve/.
63 Sister Grove Farm, "Our Story," accessed February 1, 2025, https://www.sistergrovefarm.com/story.
64 Sister Grove Farm, "Chicken," accessed February 1, 2025, https://www.sistergrovefarm.com/chicken.
65 Sister Grove Farm, "Chicken."
66 Other examples include Black Farmers' Rice Project (https://www.jubileejustice.org/sri-rice) and Agrarian Trust (https://www.agrariantrust.org/principles/).
67 Green America, "Our Story," accessed November 15, 2024, https://greenamerica.org/our-mission.
68 Green America has resources for collective work too. As we note in our later chapter on living lightly, there is a place for changing our personal habits, and that place is in the context of also working for change in institutions, communities, and broader social structures.

CHAPTER THREE: CHANGE THE RULES

1 See chapter 10 in Cynthia Moe-Lobeda, *Building a Moral Economy: Pathways for a People of Courage*, Fortress Press, 2024.
2 Anna Blaedel, "Good Friday," *enfleshed*, accessed November 15, 2024, https://enfleshed.com/project/annual/.
3 We cannot ignore that there are old rules, ancient rules that exist and are life-giving. Without romanticizing, Indigenous ways of being are often at odds with Western ways. We recommend Robin Wall Kimmerer, Randy Woodley, and geran lorraine

as scholars and practitioners to explore who are writing at the intersections of indigeneity, ecology, economy, and spiritualities.

4 "About Us," Earth Charter," accessed January 5, 2025, a http://earthcharter.org/discover/what-is-the-earth-charter/.

5 All quotations that follow are from the Earth Charter (English translation). See "Download the Charter," Earth Charter, https://earthcharter.org/read-the-earth-charter/download-the-charter/.

6 In later chapters, we talk about these two means of action or fingers on the hands of healing change—changing the narrative and amplifying the stories and voices of struggle. Here we see a beautiful illustration of how the fingers work together.

7 "ShiftUS - US Fair Share," accessed February 21, 2025, https://www.thepeoplesjusticecouncil.org/us-fair-share.

8 Natalie Lucas is the former coordinator of the People's Justice Council's US Fair Share NDC. Natalie stepped down as the coordinator and handed the leadership to two other youth leaders, Mareshah Malcom and Analyah Schlaeger Dos Santos. Mareshah Malcom is the program director for the People's Justice Council and Alabama Interfaith Power and Light. Analyah Schlaeger Dos Santos is the youth program director at Minnesota Interfaith Power and Light. Natalie is still actively involved in US Fair Shares and the Shift US campaign.

9 See https://thirdact.org/. They let people who are under the age of sixty join them too. abby has led chants and songs with the Chicago chapter.

10 "Who We Are," about page, Third Act, accessed February 1, 2025, https://thirdact.org/about/who-we-are/.

11 Descriptions for each of these principles are available on the website for Third Act.

12 See https://fossilfueltreaty.org/.

13 Read more here: Fossil Fuel Non-Proliferation Treaty, accessed February 1, 2025, https://fossilfueltreaty.org/.

14 Fossil Fuel Non-Proliferation Treaty.

15 As noted, Michael is the executive director of Alabama Interfaith Power and Light. abby interned with, worked for, volunteers with, and supports Faith in Place, the Illinois affiliate.

16 Martin Niemöller, "First They Came," Holocaust Memorial Day Trust, accessed February 21, 2025, https://hmd.org.uk/resource/first-they-came-by-pastor-martin-niemoller/.

CHAPTER FOUR: MOVE THE MONEY

1 "Divestment" has been described as both a movement and a tactic (that is, a part of a larger strategy). In this paper, I use it as a "tactic," understanding that both critics and proponents of divestment suggest that it is most effective as part of a larger strategy. In fact, as I explore more below, divestment is a tactic used to draw international attention to and conversation about a particular injustice—not always as the dismantling of the injustice itself.

2 Leonard Vignola, *Strategic Divestment* (ACACOM, 1974) xi.

3 Based on Desmond Tutu's words: "My humanity is bound up in yours and we can only be human together." Allister Sparks and Mpho Tutu, *Tutu Authorized* (PQ Blackwell Limited, 2011), 339. Similar language is lifted up in the WECAN divestment report (See Mary-Elizabeth Estrada et al., "Gendered and Racial Impacts of the Fossil Fuel Industry in North America and Complicit Financial Institutions," Women's Earth and Climate Action Network (WECAN) International, 2024, https://www.wecaninternational.org/divestment-report.

4 Boycotting has been used for centuries (more below), but the practice of disinvestment and divestment as a particular boycott of stocks was first used in response to a call by South Africans. Divestment as an economic strategy was first reported on in the 1970s but used before the 1950s. Vignola, *Strategic Divestment*, 2.

5 Desmond Tutu, "We Need an Apartheid-Style Boycott to Save the Planet," *The Guardian*, April 10, 2014, https://www.theguardian.com/commentisfree/2014/apr/10/divest-fossil-fuels-climate-change-keystone-xl. Bill McKibben also included in his interview that when it was time for 350.org to consider divestment as a tactic, McKibben wrote to Tutu for permission, a clear indication that Tutu's leadership and authority was respected and acknowledged. Another movement that relied on Tutu's authority, as will be seen below, is the Boycott, Divestment, and Sanctions (BDS) movement in which Palestinians and their allies call for international solidarity through BDS in order to make the world pay attention to the unjust reality of Palestinians under the state of Israel. Tutu's role as a faith leader in South Africa is also explored in more detail in John Allen, *Desmond Tutu: Rabble-rouser for Peace* (Free Press, 2006).

6 Charles W. Powers, "Case Studies in the Pursuit of Ethical Criteria for Social Investment Activity," in *People / Profits: The Ethics of Investments*, ed. Charles W. Powers (Council on Religion and International Affairs, 1972), 51. Powers is one of the students at Union Theological Seminary who successfully organized the Banks Campaign, the movement to get ten US banks to stop lending money to the apartheid government in South Africa (see Robert Ross, "Moving the Mountain of Apartheid: The 1966–1969 Banks Campaign and the Rise of Economic Tools to End South African Apartheid," *Safundi: The Journal of South African and American Studies*, 22:4, 303–329).

7 Vernon S. Broyles III, "When Stewardship Is an Act of Public Witness," Presbyterian Church (USA), published September 10, 2014, https://www.pcusaPC(USA).org/news/2014/9/10/when-stewardship-act-public-witness/.

8 David Feldman, "Boycotts: From the American Revolution to BDS," in *Boycotts Past and Present: From the American Revolution to the Campaign to Boycott Israel*, ed. David Feldman (Palgrave Macmillan, 2019), 7–9.

9 Feldman., 4.

10 Feldman, 15.

11 Feldman, 12.

12 On the history webpage of the Interfaith Center for Corporate Responsibility, they describe their founding as a commitment "to employ a more direct strategy by using their financial stake in some of the world's most powerful companies to promote

a corporate response to the human rights abuses occurring under the racist apartheid system in South Africa." ("Our Origin Story," Interfaith Center for Corporate Responsibility, accessed February 9, 2024, https://www.iccr.org/mission-history/.)

13 Andrew Behar, *The Shareholder Action Guide: Unleash Your Hidden Powers to Hold Corporations Accountable* (Berrett-Koehler Publishers, 2016), 18–19. Resolutions by shareholders need to be voted on by shareholders and only resolutions that receive the majority of votes change policy.

14 Behar, 19.

15 Behar, 125.

16 Powers, *Social Responsibility and Investment*, 88.

17 Bill McKibben, "Global Warming's Terrifying New Math," *Rolling Stone*, last modified July 19, 2011, https://www.rollingstone.com/politics/politics-news/global-warmings-terrifying-new-math-188550/.

18 Making a case for how the fossil fuel industry causes climate change is outside the scope of this paper, but it is the general scientific consensus that climate change is real, is caused by humans and, among other things, worsened by our burning of fossil fuels.

19 McKibben has written at least two follow-up articles to reach target audiences. He wrote specifically for a Christian audience, referencing his baptism as a Presbyterian and background as a Methodist Sunday School teacher. (See McKibben's "Playing Offense: It's Time to Divest from the Fossil Fuel Industry," 2013.) He's also returned to popular news media, beating the drum for more college campuses and universities to join the movement. (See McKibben, "It's Time to Stop Investing in the Fossil Fuel Industry," 2013).

20 Bill McKibben, "It's Time to Stop Investing in the Fossil Fuel Industry," *The Guardian*, May 30, 2013, https://www.theguardian.com/commentisfree/2013/may/30/fossil-fuel-divestment-climate-change.

21 Personal Communication with Jim Antal, 2017.

22 Personal communication with Dan Terpstra, 2021.

23 Jenny Phillips has since left the leadership of Fossil Free UMC, moving onto the board of West Path—the pension board of the United Methodist Church—and then into the denomination's environmental ministry staff position. Because of the variety of her roles and the fiduciary responsibility she has to the UMC, she spoke with me on the condition that any phrasing that risked that responsibility was attributed to me.

24 Personal communication with Fletcher Harper, 2017.

25 From my interview with Jim Antal: "Early on in the divestment movement at Harvard [there were those] who were very vocal about Harvard's moral responsibility to divest and then 2014 there were 120 faculty who wrote an articulate letter to the president urging the president to listen to the students[.] President Drew Faust wrote a statement that was embarrassing, in which she made her case that the investment of Harvard must be blind to moral concerns. It's so dramatic—the letter from the faculty was so articulate and the letter form the president was shockingly immoral. By the end of 2015 rumor had it that Harvard's

investment managers had purged Harvard of fossil fuels but didn't want to put in some sort of future bind that they had paid attention to the wishes of students and faculty." (More information about Harvard's divestment is available here at Oliver Milman, "Harvard University Pausing Investments," *The Guardian*, April 27, 2017, https://www.theguardian.com/environment/2017/apr/27/harvard-university-pausing-investments-in-some-fossil-fuels.

26 Naomi Klein, *This Changes Everything* (New York: Simon and Schuster 2014), 354.

27 Laurel Kearns, "The Role of Religions in Activism," in *The Oxford Handbook on ClimateChange and Society*, edited by John Dryzek, Richard Norgaard and David Schlosberg, (Oxford University Press, 2011).

28 See the resolution here: https://www.sneucc.org/files/files/documentsenvironmentalministries/ucc_fossilfueldivestmentresolution_asvoted_july1_2013.pdf.

29 Essentially this means that ten other regions voted in solidarity with the Massachusetts Conference. (The United Church of Christ calls each region a conference).

30 Personal communication, 2017.

31 Nepstad, 5.

32 See Rich Copley, New Faith-Based Investment Chair Focusing On People," Presbyterian News Service, November 4, 2020, https://pcusa.org/news-storytelling/news/new-faith-based-investment-chair-focusing-people.

33 These actions happened in a larger context of growing concern by the PC(USA) about climate change around the world, as evidenced by an increased number of policies, teachings, and conversations at all levels of the denomination.

34 The Carbon Underground 200 is also referenced in the first overture that Fossil Free PC(USA) wrote. It is "an annually updated listing of the top 100 public coal companies globally and the top 100 public oil and gas companies globally, ranked by the potential carbon emissions content of their reported reserves." ("Thousands join Fossil Fuel Divestment Campaign Global Day of Action," Fossil Free UK, February 13, 2015, https://gofossilfree.org/uk/press-release/thousands-join-fossil-fuel-divestment-campaign-global-day-of-action/.)

35 Nepstad, 9.

36 Steven Feit, Amanda Kistler, Carroll Muffett, and Lisa Hamilton. "Trillion Dollar Transformation: Fiduciary Duty, Divestment, and Fossil Fuels in an Era of Climate Risk." Center for International Environmental Law, December 2016, http://www.ciel.org/wp-content/uploads/2016/12/Trillion-Dollar-Transformation-CIEL.pdf.

37 Feit et al., 3.

38 In a later chapter, we talk about the call to live lightly on the planet with our whole selves and how this is a necessary act for people with structural privilege in the world. Divestment is one way to live lightly (another reason why this chapter is so long—it's a good example of how many different fingers on the hands of action are connected!).

39 For example, early proponents of divestment as a tactic in social movements were Leon Sullivan, a Black American pastor, and Archbishop Desmond Tutu; each

called upon the United States as a historically wealthy nation (wealth built on the exploitation of people of color and Black people in particular) to use its wealth to be in solidarity with Black South Africans who did not have the economic privilege or infrastructure to divest from their own oppression.

40 "Dirty Pearls: Exposing Shell's Hidden Legacy of Climate Change Accountability, 1970–1990," accessed February 23, 2025, https://changerism.com/portfolio/dirty-pearls-exposing-shells-hidden-legacy-of-climate-change-accountability-1970-1990/.

CHAPTER FIVE: GROW THE BIGGER WE

1 "Life is life-ing" is a phrase that means that life is happening in the full spectrum of experiences.

2 Melanie Harris, *Ecowomanism* (Orbis Books, 2017) 28.

3 Melanie Harris, "Ecowomanism: Black Women, Religion, and the Environment," *The Black Scholar* 46, no. 3 (2016), 28.

4 Harris, 30.

5 I'm indebted here to conversations with my colleague Dawrell Rich.

CHAPTER SIX: RESIST THE WRONG

1 Fletcher Harper, "Important Update: Faiths for Climate Justice," GreenFaith, accessed January 6, 2025, https://greenfaith.org/important-update-faiths-for-climate-justice/. We're intentionally being a little vague about the details of these actions because the stakes are high for these people of faith. They could lose much more than their phones. Our vague descriptions are intended to protect them and their communities, while also pointing to their faithful courage. Our deep thanks to Meryne Warah, global organizer at GreenFaith, for reviewing this chapter and advising us.

2 Fletcher Harper and Justine Hughes, "As If Nothing Is Sacred," GreenFaith, last updated November 9, 2023, https://greenfaith.org/as-if-nothing-is-sacred/.

3 "Take Action: Support the Tanzania Ten, Interfaith Solidarity Fast for Climate Justice," GreenFaith, accessed July 18, 2024, https://greenfaith.org/interfaith-solidarity-fast/.

4 "Open Letter to Samia Suluhu Hassan, President of Tanzania and Patrick Pouyanné, CEO, TotalEnergies," GreenFaith, accessed July 18, 2024, https://greenfaith.org/people-of-faith-to-samia-suluhu-hassan-and-patrick-pouyanne/.

5 For how conservative Christians have sought to debunk climate science, see E. Calvin Beisner, Paul K. Driessen, Ross McKitrick, and Roy W. Spencer, "A Call to Truth, Prudence, and Protection of the Poor: An Evangelical Response to Global Warming," Cornwall Alliance for the Stewardship of Creation, accessed August 18, 2024, https://www.cornwallalliance.org/docs/a-call-to-truth-prudence-and-protection-of-the-poor.pdf

6 Stephane Hallegatte et al., *Shock Waves: Managing the Impacts of Climate Change on Poverty*, Climate Change and Development Series, World Bank, 2016.

7 Kate Abnett and Alison Withers, "2024 Could Be World's Hottest Year as June Breaks Records," *Reuters*, July 8, 2024, https://www.reuters.com/business/environment/2024-could-be-worlds-hottest-year-june-breaks-records-20240–70–8. Anecdotally, when abby was writing her dissertation on faith-based social movements around climate change (which took five years), she had to keep updating the date of the hottest year on record. Each year was hotter than the last.
8 David Wallace-Wells, "The Tragedy at This Year's Hajj Is Just the Beginning," *New York Times*, July 10, 2024, https://www.nytimes.com/2024/07/10/opinion/hajj-heat-climate-death.html.
9 Debra Rienstra, "Why Does Religion Get in the Way of Climate Action?", last updated February 18, 2023, https://debrarienstra.com/why-does-religion-get-in-the-way-of-climate-action/.
10 Dayenu, "What We Do," accessed August 18, 2024, https://dayenu.org/what-we-do/.
11 Talitha Amadea Aho, *In Deep Waters: Spiritual Care for Young People in a Climate Crisis* (Fortress Press, 2022).
12 Sallie McFague, *The Body of God: An Ecological Theology*, (Fortress Press, 1993).
13 Meg Duff, "Christians Have Climate Grief. How Can Churches Support Them?" *Sojourners*, June 12, 2024, https://sojo.net/articles/christians-have-climate-grief-how-can-churches-support-them.
14 Personal communication with author, 2021.
15 Nathaniel Rich, *Losing Earth: The Decade We Almost Stopped Climate Change*, last modified August 1, 2018, https://www.nytimes.com/interactive/2018/08/01/magazine/climate-change-losing-earth.html.
16 The English translation of the Hebrew word *adamah* in Genesis 2:7, often translated as "dust."

CHAPTER SEVEN: CHANGE THE STORY

1 Sasha-Ann Simons, host, *1A*, podcast, "To Be in a Rage, Almost All The Time," *NPR*, June 1, 2020, https://www.npr.org/2020/06/01/867153918/-to-be-in-a-rage-almost-all-the-time.
2 DiAngelo, 20.
3 Richard Delgado and Jean Stefancic, *Critical Race Theory: An Introduction*, 3rd ed. (NYU Press, 2017), 21.
4 Richard Rothstein, *The Color of Law: A Forgotten History of How Our Government Segregated America* (Liveright, 2018), 52.
5 Rothstein, 52.
6 Rothstein, 54.
7 Cedric J. Robinson, *Black Marxism: The Making of the Black Radical Tradition*, 3rd edition (University of North Carolina Press, 1983), 189.
8 Nedra Rhone, "50 Years After Fair Housing Act, Residential Segregation in Atlanta Remains High," *The Atlanta Journal Constitution*, July 6, 2022, https://www.ajc.com/life/opinion-50-years-after-fair-housing-act-residential-segregation-in-atlanta-remains-high/5H7KZHE4LNBX5BGIGNH2LSHHR4/#.

9 Robert Bullard, *Dumping in Dixie: Race, Class, and Environmental Quality* (Westview Press, 1990), 91.
10 *Merriam-Webster Dictionary,* "white supremacy," accessed February 23, 2025, https://www.merriam-webster.com/dictionary/white%20supremacy.
11 See "UDC Information," Historic Bleak House, accessed February 23, 2025, http://www.bleakhouseudc89.org/udc_info.htm.
12 Guthrie, 213.
13 Alonso Johnson and Mark Koenig, "Antiracism Next Steps," presentation at Compassion, Peace, and Justice Day, Washington, DC, April 21, 2017.
14 Brian Kahn, "The World Passes 400 PPM Threshold. Permanently," *Climate Central*, published September 27, 2016, http://www.climatecentral.org/news/world-passes-400-ppm-threshold-permanently-20738.
15 See more on climate debt in Klein, 408–410.
16 Learn more about Climate Reparations at Sara Schonhardt, "5 Things to Know About Climate Reparations," *Scientific American*, published October 25, 2022, https://www.scientificamerican.com/article/5-things-to-know-about-climate-reparations/.
17 Marvin H. Pope, *Job: The Anchor Bible* (Doubleday, 1973), 24.
18 Pope, 24.
19 Hiebert, 6.

CHAPTER EIGHT: LIVE LIGHTLY

1 United Nations, "Greenwashing—the Deceptive Tactics Behind Environmental Claims," accessed February 1, 2025, https://www.un.org/en/climatechange/science/climate-issues/greenwashing.
2 Tess Riley, "Just 100 Companies Responsible for 71% of Global Emissions, Study Says," *The Guardian*, July 10, 2017, https://www.theguardian.com/sustainable-business/2017/jul/10/100-fossil-fuel-companies-investors-responsible-71-global-emissions-cdp-study-climate-change.
3 Carbon footprint tracking was invented by British Petroleum as a way to make dealing with climate change the consumer's problem. It's a ploy to make us think "Oh, if I just fly less or eat less meat, I will save the world." But individuals cannot change laws and infrastructure the way companies and governments can. If we are supposed to track our own footprints, those footprints should be placed alongside the ginormous footprints of fossil fuel companies. See more at Pam Reynolds, "The Truth About Carbon Footprints," Conservation Law Foundation, March 18, 2024, https://www.clf.org/blog/the-truth-about-carbon-footprints.
4 Fact Sheet: Climate Change, United Nations Sustainable Transport Conference, October 14–16, 2021, https://www.un.org/sites/un2.un.org/files/media_gstc/FACT_SHEET_Climate_Change.pdf.
5 Ben Welle et al., "Post-Pandemic, Public Transport Needs to Get Back on Track to Meet Global Climate Goals," World Resources Institute, December 14, 2023, https://www.wri.org/insights/current-state-of-public-transport-climate-goals.
6 Welle. See also this website for a global climate map of transportation: https://outlook.transformative-mobility.org/.

7 TransitCenter, "The State of Transit Equity: Metro Chicago," accessed February 1, 2025, https://transitcenter.org/wp-content/uploads/2021/06/ChicagoFactSheet.pdf
8 I know: industrial agriculture. See the next little section. Still, there is something about those cornfields.
9 I lived for two years on a regenerative farm where I cared for chickens, cows, soil, and people (Sister Grove, which we mentioned in an earlier chapter). I learned a ton about how industrial farming is wrecking the planet and people, and I also learned that I can't eat even loved-on animals. I don't like the taste, and animals deserve to be eaten by people who will enjoy them. A later book in this same series is about food justice, and it will go more deeply into these relationships.
10 One of the things we have to consider when we talk about eating lower on the food chain includes travel distances for food. Soybeans are great for tofu, but if it's being shipped across the world, it's really better to eat locally sourced beef from down the road.
11 More at "Home Weatherization," Illinois Department of Commerce and Economic Opportunity, accessed February 1, 2025, https://dceo.illinois.gov/communityservices/homeweatherization.html.
12 Sort of like a stop sign campaign in organizing. See George Goehl, "Back to Basics: The Fundamentals of Community Organizing," *Convergence*, May 29, 2024, https://convergencemag.com/articles/back-to-basics-the-fundamentals-of-community-organizing/.
13 "Mission," about page, Beyond Plastics, https://www.beyondplastics.org/about.

CHAPTER NINE: LISTEN AND AMPLIFY

1 "The Movement for Black Lives," home page, M4BL, https://m4bl.org/.

BIBLIOGRAPHY

A Wisdom Keepers Delegation. "Who We Are." Accessed January 4, 2025. https://www.awisdomkeepersdelegation.org/.

Abnett, Kate, and Alison Withers. "2024 Could Be World's Hottest Year As June Breaks Records." *Reuters*. Published July 8, 2024. https://www.reuters.com/business/environment/2024-could-be-worlds-hottest-year-june-breaks-records-20240–70–8.

Aho, Talitha Amadea. *In Deep Waters: Spiritual Care for Young People in a Climate Crisis*. Fortress Press, 2022.

Allen, John. *Desmond Tutu: Rabble-Rouser for Peace*. Free Press, 2006.

Atkinson, David. *Renewing the Face of the Earth: A Theological and Pastoral Response to Climate Change*. Canterbury Press, 2008.

Baker-Fletcher, Karen. "To Live as People of Dust and Spirit." In *Earth and Word*, edited by David Rhoads. Continuum, 2007.

Bauman, Whitney A. "What's Left (Out) of the Lynn White Narrative?" In *Religion and Ecological Crisis: The "Lynn White Thesis" at Fifty*, edited by Todd LeVasseur and Anna Peterson. Routledge, 2017.

Behar, Andrew. *The Shareholder Action Guide: Unleash Your Hidden Powers to Hold Corporations Accountable*. Berrett-Koehler Publishers, 2016.

Beisner, E. Calvin, Paul K. Driessen, Ross McKitrick, and Roy W. Spencer. "A Call to Truth, Prudence, and Protection of the Poor: An Evangelical Response to Global Warming." Cornwall Alliance for the Stewardship of Creation. Accessed August 18, 2024. https://www.cornwallalliance.org/docs/a-call-to-truth-prudence-and-protection-of-the-poor.pdf.

Bernstein, Larry, Peter Bosch, Osvaldo Canziani, et al. "Climate Change 2007: Synthesis Report. Intergovernmental Panel on Climate Change Plenary XXVII." Published November 12, 2007. https://www.ipcc.ch/pdf/assessment-report/ar4/syr/ar4_syr.pdf.

Blaedel, Anna. "Good Friday." *enfleshed*. Accessed November 15, 2024. https://enfleshed.com/project/annual/.

Broyles III, Vernon S. "When Stewardship Is an Act of Public Witness." Presbyterian Church (USA). Last modified September 10, 2014. https://www.pcusaPC(USA).org/news/2014/9/10/when-stewardship-act-public-witness/.

Bullard, Robert. *Dumping in Dixie: Race, Class, and Environmental Quality*. Westview Press, 1990.

Case, Liz. "In the Beginning: How Arrayed Got Started." *Arrayed*, accessed February 1, 2025. https://www.shoparrayed.com/blogs/news/how-arrayed-got-started.

Changerism. "Dirty pearls: exposing Shell's hidden legacy of climate change accountability, 1970–1990," accessed February 23, 2025, https://changerism.com/portfolio/dirty-pearls-exposing-shells-hidden-legacy-of-climate-change-accountability-1970-1990/.

Cone, James. "Whose Earth Is It Anyway?" In *Earth Habitat*. Edited by Dieter Hessel and Larry Rasmussen. Fortress Press, 2001.

Confederate Memorial Hall (Bleak House). "UDC Information." Accessed February 23, 2025. http://www.bleakhouseudc89.org/udc_info.htm.

Connolly, William E. *Facing the Planetary: Entangled Humanism and the Politics of Swarming*. Duke University Press, 2017.

Copley, Rich. "New Faith-Based Investment Chair Focusing On People." Presbyterian News Service. Accessed November 4, 2020. https://pcusa.org/news-storytelling/news/new-faith-based-investment-chair-focusing-people.

Cunningham, Liz. *Ocean Country*. North Atlantic Books, 2015.

Darymple, Amy. "Pipeline Route Plan First Called for Crossing North of Bismarck." *The Bismarck Tribune*, last modified August 18, 2016. http://bismarcktribune.com/news/state-and-regional/pipeline-route-plan-first-called-for-crossing-north-of-bismarck/article_64d053e48–a1a-5198-a1dd-498d386c933c.html.

Delgado, Richard and Jean Stefancic. *Critical Race Theory: An Introduction*. 3rd ed. NYU Press, 2017.

DiAngelo, Robin J. *White Fragility: Why It's So Hard for White People to Talk About Racism*. Beacon Press, 2018.

Duchrow, Ulrich. *Alternatives to Global Capitalism: Drawn from Biblical History, Designed for Political Action*. Kairos Europa, 1997.

Duff, Meg. "Christians Have 'Climate Grief.' How Can Churches Support Them?" *Sojourners*. Published June 12, 2024. https://sojo.net/articles/christians-have-climate-grief-how-can-churches-support-them.

Earth Charter. "Download the Charter." Accessed January 5, 2025. https://earthcharter.org/read-the-earth-charter/download-the-charter/.

Earth Equity Advisors. Home page. Accessed January 4, 2025. https://www.earthequityadvisors.com/.

Eligon, John. "A Question of Environmental Racism in Flint." *New York Times*, published January 21, 2016. https://www.nytimes.com/2016/01/22/us/a-question-of-environmental-racism-in-flint.html?_r=0.

Ernst-Habib, Margit. "A Conversation with Twentieth Century Confessions." In *Conversations with the Confessions*. Edited by Joseph D. Small. Geneva Press, 2005.

Estrada, Mary-Elizabeth, Livia Charles, Allison Fabrizio, and Osprey Orielle Lake. "Gendered and Racial Impacts of the Fossil Fuel Industry in North America and Complicit Financial Institutions." Women's Earth and Climate Action Network (WECAN) International. Accessed March 17, 2025. https://www.wecaninternational.org/divestment-report.

Feit, Steven, Amanda Kistler, Carroll Muffett, and Lisa Hamilton. "Trillion Dollar Transformation: Fiduciary Duty, Divestment, and Fossil Fuels in an Era of Climate Risk." Center for International Environmental Law. Published December 2016. http://www.ciel.org/wp-content/uploads/2016/12/Trillion-Dollar-Transformation-CIEL.pdf.

Feldman, David. "Boycotts: From the American Revolution to BDS." In *Boycotts Past and Present: From the American Revolution to the Campaign to Boycott Israel*. Edited by David Feldman. Palgrave Macmillan, 2019.

Fossil Free UK. "Thousands Join Fossil Fuel Divestment Campaign Global Day of Action." Published February 13, 2015. https://gofossilfree.org/uk/press-release/thousands-join-fossil-fuel-divestment-campaign-global-day-of-action/.

Fossil Fuel Non-Proliferation Treaty. Home page. Accessed February 1, 2025. https://fossilfueltreaty.org/.

Goehl, George. "Back to Basics: The Fundamentals of Community Organizing." *Convergence*, published May 29, 2024. https://convergencemag.com/articles/back-to-basics-the-fundamentals-of-community-organizing/.

Green America. "Our Story." Accessed November 15, 2024, https://greenamerica.org/our-mission.

GreenFaith. "Open Letter to Samia Suluhu Hassan, President of Tanzania and Patrick Pouyanné, CEO, TotalEnergies." Accessed July 18, 2024. https://greenfaith.org/people-of-faith-to-samia-suluhu-hassan-and-patrick-pouyanne/.

GreenFaith. "Take Action: Support the Tanzania Ten, Interfaith Solidarity Fast for Climate Justice." Accessed July 18, 2024. https://greenfaith.org/interfaith-solidarity-fast/.

Guthrie, Shirley. *Christian Doctrine*. Westminster John Knox Press, 1994.

Hallegatte, Stephane, Mook Bangalore, Laura Bonzanigo, et al. *Shock Waves: Managing the Impacts of Climate Change on Poverty*. Climate Change and Development Series. World Bank. 2016

Harper, Fletcher, and Justine Hughes. "As If Nothing Is Sacred." GreenFaith. Last updated November 9, 2023. https://greenfaith.org/as-if-nothing-is-sacred/.

Harper, Fletcher. "Important Update: Faiths for Climate Justice." GreenFaith. Accessed January 6, 2025. https://greenfaith.org/important-update-faiths-for-climate-justice/.

Harris, Melanie. "Ecowomanism: Black Women, Religion, and the Environment." *The Black Scholar* 46, no. 3 (2016).

Harris, Melanie. *Ecowomanism*. Orbis Books, 2017.

Hiebert, Theodore. *The Yahwist Landscape*. Oxford University Press, 1996.

Hughes, Langston. "I, Too." *Poetry Foundation*. https://www.poetryfoundation.org/poems/47558/i-too.

Illinois Department of Commerce and Economic Opportunity. "Home Weatherization." Accessed February 1, 2025. https://dceo.illinois.gov/communityservices/homeweatherization.html.

Interfaith Center for Corporate Responsibility. "Our Origin Story." Accessed February 9, 2024. https://www.iccr.org/mission-history/.

Johnson, Alonso and Mark Koenig. "Antiracism Next Steps." Presentation. Compassion, Peace, and Justice Day. Washington, DC. April 21, 2017.

Kahn, Brian. "The World Passes 400 PPM Threshold. Permanently." *Climate Central*, published September 27, 2016. http://www.climatecentral.org/news/world-passes-400-ppm-threshold-permanently-20738.

Kearns, Laurel. "The Role of Religions in Activism." In *The Oxford Handbook on Climate Change and Society*. Edited by John Dryzek, Richard Norgaard, and David Schlosberg. Oxford University Press, 2011.

Klein, Naomi. *This Changes Everything*. Simon and Schuster, 2014.

Koinonia Farm. "About Us." Accessed March 17, 2025. https://www.koinoniafarm.org/about-us.

Kruse, Kevin. *One Nation Under God: How Corporate America Invented Christian America*. Basic Books, 2015.

LaCocque, André. *Ruth: A Continental Commentary*. Fortress Press, 2004.

Lang, Olivia. "Maldives Leader in Climate Change Stunt." *BBC*, published October 17, 2009. http://news.bbc.co.uk/2/hi/south_asia/8312320.stm.

Maathai, Wangari. *Replenishing the Earth: Spiritual Values for Healing Ourselves and the World*. Double Day, 2010.

Martínez de Anguita, Pablo. *Environmental Solidarity: How Religions Can Sustain Sustainability*. Routledge, 2012.

Martin-Schramm, James B. *Climate Justice: Ethics, Energy, and Public Policy*. Fortress Press, 2010.

McFague, Sallie. *The Body of God: An Ecological Theology*. Fortress Press, 1993.

McKibben, Bill. "Global Warming's Terrifying New Math." *Rolling Stone*, last modified July 19, 2011. https://www.rollingstone.com/politics/politics-news/global-warmings-terrifying-new-math-188550/.

———. "Playing Offense: It's Time To Divest From The Oil Industry." *The Christian Century*. January 9, 2013. https://www.christiancentury.org/article/20121–2/playing-offense.

Melillo, Jerry, Terese (T.C.) Richmond, and Gary W. Yohe, eds. *Highlights of Climate Change Impacts in the United States: The Third National Climate Assessment*. U.S. Global Change Research Program, 2014.

Merriam-Webster Dictionary. "White supremacy." Accessed February 23, 2025. https://www.merriam-webster.com/dictionary/white%20supremacy.

Migliore, Daniel L. *Faith Seeking Understanding: An Introduction to Christian Theology*. Eerdmans, 1991.

Miller, Kelsey. "The Triple Bottom Line: What It Is & Why It's Important." *Business Insights*, published December 8, 2020. https://online.hbs.edu/blog/post/what-is-the-triple-bottom-line.

Milman, Oliver. "Harvard University Pausing Investments." *The Guardian*, published April 27, 2017. https://www.theguardian.com/environment/2017/apr/27/harvard-university-pausing-investments-in-some-fossil-fuels.

Moe-Lobeda, Cynthia. *Building a Moral Economy: Pathways for People of Courage*. Fortress Press, 2024.

Niemöller, Martin. "First They Came." Holocaust Memorial Day Trust. Accessed February 21, 2025. https://hmd.org.uk/resource/first-they-came-by-pastor-martin-niemoller/.

Ogunyinka, Oyinkansola. "Think Before You Take: 6 AAVE Terms Brands Should Use More Responsibly." Writer's Room. Accessed February 1, 2025. https://writer.com/blog/aave-terms/.

Pope, Marvin H. *Job: The Anchor Bible*. Doubleday, 1973.

Powers, Charles W. "Case Studies in the Pursuit of Ethical Criteria for Social Investment Activity." In *People/Profits: The Ethics of Investments*. Edited by Charles W. Powers. New York: Council on Religion and International Affairs, 1972.

Rao, Sairo. Accessed September 29, 2024. instagram.com/p/DAhdZHGsB4G/?img_index=10 (since deleted).

Rasmussen, Larry. "Eco-Justice: Church and Community Together." In *Earth Habitat*. Edited by Dieter Hessel and Larry Rasmussen. Fortress Press, 2001.

Reynolds, Pam. "The Truth About Carbon Footprints." Conservation Law Foundation. Published March 18, 2024. https://www.clf.org/blog/the-truth-about-carbon-footprints.

Rhone, Nedra. "50 Years After Fair Housing Act, Residential Segregation in Atlanta Remains High." *The Atlanta Journal Constitution*, published July 6, 2022. https://www.ajc.com/life/opinion-50-years-after-fair-housing-act-residential-segregation-in-atlanta-remains-high/5H-7KZHE4LNBX5BGIGNH2LSHHR4/#

Rich, Nathaniel. "Losing Earth: The Decade We Almost Stopped Climate Change." *New York Times*, published August 1, 2018. https://www.nytimes.com/interactive/2018/08/01/magazine/climate-change-losing-earth.html.

Rienstra, Debra. "Why Does Religion Get in the Way of Climate Action?" *Debra Rienstra* (blog), published February 18, 2023. https://debrarienstra.com/why-does-religion-get-in-the-way-of-climate-action/.

Riley, Tess. "Just 100 Companies Responsible for 71% of Global Emissions, Study Says." *The Guardian*. July 10, 2017. https://www.theguardian.com/sustainable-business/2017/jul/10/100-fossil-fuel-companies-investors-responsible-71-global-emissions-cdp-study-climate-change.

Robinson, Cedric J. *Black Marxism: The Making of the Black Radical Tradition*. 3rd edition. University of North Carolina Press, 1983.

Ross, Robert. "Moving the Mountain of Apartheid: The 1966–1969 Banks Campaign and the Rise of Economic Tools to End South African Apartheid." *Safundi: The Journal of South African and American Studies*, 22:4, 303–329.

Rothstein, Richard. *The Color of Law: A Forgotten History of How Our Government Segregated America*. Norton, 2018.

Rowe, Terra. *Toward a Better Worldliness: Ecology, Economy, and the Protestant Tradition*. Fortress Press, 2017.

Sasha-Ann Simons, host. *1A*, podcast. "To Be In a Rage, Almost All The Time." *NPR*. Published June 1, 2020. https://www.npr.org/2020/06/01/867153918/-to-be-in-a-rage-almost-all-the-time.

Schonhardt, Sara. "5 Things to Know about Climate Reparations." *Scientific American*, published October 25, 2022. https://www.scientificamerican.com/article/5-things-to-know-about-climate-reparations/.

Sister Grove Farm. "Our Story." Accessed February 1, 2025. https://www.sister-grovefarm.com/story.

Skelton, Renee, and Vernice Miller. "The Environmental Justice Movement." Natural Resources Defense Council. Last modified March 17, 2016. https://www.nrdc.org/stories/environmental-justice-movement.

Sparks, Allister and Mpho Tutu. *Tutu Authorized*. PQ Blackwell Limited, 2011.

The People's Justice Council. "ShiftUS - US Fair Share." Accessed February 21, 2025. https://www.thepeoplesjusticecouncil.org/us-fair-share.

Third Act. "Who We Are." Accessed February 1, 2025. https://thirdact.org/about/who-we-are/.

Tierney, Kathleen. "Foreshadowing Katrina: Recent Sociological Contributions to Vulnerability Science." *Contemporary Sociology* 35, no. 3 (May 2006): 207.

Toner, Kathleen. "By Nourishing Plants, You're Nourishing Community." *CNN*, last modified on March 21, 2016. http://www.cnn.com/2015/09/24/us/cnn-heroes-joyner/.

TransitCenter. "The State of Transit Equity: Metro Chicago." Accessed February 1, 2025. https://transitcenter.org/wp-content/uploads/2021/06/ChicagoFactSheet.pdf.

Trentmann, Frank. "Consumer Boycotts in Modern History: States, Moral Boundaries, and Political Action." In *Boycotts Past and Present: From the American Revolution to the Campaign to Boycott Israel*. Edited by David Feldman. Palgrave Macmillan, 2019.

Tutu, Desmond. "We Need an Apartheid-Style Boycott to Save the Planet." *The Guardian*, published April 10, 2014. https://www.theguardian.com/commentisfree/2014/apr/10/divest-fossil-fuels-climate-change-keystone-xl.

United Nations. "Fact Sheet: Climate Change." October 14–16, 2021. https://www.un.org/sites/un2.un.org/files/media_gstc/FACT_SHEET_Climate_Change.pdf

United Nations: Climate Action. "Greenwashing—the Deceptive Tactics Behind Environmental Claims." Accessed February 1, 2025. https://www.un.org/en/climatechange/science/climate-issues/greenwashing.

Vignola, Leonard. *Strategic Divestment*. ACACOM, 1974.

Wallace-Wells, David. "The Tragedy at This Year's Hajj Is Just the Beginning." *New York Times*, published July 10, 2024. https://www.nytimes.com/2024/07/10/opinion/hajj-heat-climate-death.html.

Welle, Ben, Anna Kustar, Thet Hein Tun, and Cristina Albuquerque. "Post-Pandemic, Public Transport Needs to Get Back on Track to Meet Global Climate Goals." World Resources Institute. Published December 14, 2023. https://www.wri.org/insights/current-state-of-public-transport-climate-goals.

White, Carol Wayne. "Religious Naturalism: A New Ideal in Africana Religious Thought." Paper presented at the Transdisciplinary Theological Colloquium at Drew University in Madison, New Jersey, March 24–26, 2017.

INDEX